D1104506

SI units

Physical quantity	Old unit	Value in SI units
energy	calorie (thermochemical)	4·184 J (joule)
	*electronvolt—eV	$1·602 \times 10^{-19}$ J
	*electronvolt per molecule	96·48 kJ mol^{-1}
	erg	10^{-7} J
	*wave number—cm^{-1}	$1·986 \times 10^{-23}$ J
entropy (S)	eu = cal g^{-1} °C^{-1}	4184 J kg^{-1} K^{-1}
force	dyne	10^{-5} N (newton)
pressure (P)	atmosphere	$1·013 \times 10^5$ Pa (pascal), or N m^{-2}
	torr = mmHg	133·3 Pa
dipole moment (μ)	debye—D	$3·334 \times 10^{-30}$ C m
magnetic flux density (H)	*gauss—G	10^{-4} T (tesla)
frequency (v)	cycle per second	1 Hz (hertz)
relative permittivity (ε)	dielectric constant	1
temperature (T)	*°C and °K	1 K (kelvin); 0 °C = 273·2 K

(* indicates permitted non-SI unit)

Multiples of the base units are illustrated by length

fraction	10^9	10^6	10^3	1	(10^{-2})	10^{-3}	10^{-6}	10^{-9}	(10^{-10})	10^{-12}
prefix	giga-	mega-	kilo-	metre	(centi-)	milli-	micro-	nano-	(*ångstrom)	pico-
unit	Gm	Mm	km	m	(cm)	mm	μm	nm	(*Å)	pm

The fundamental constants

Avogadro constant	L or N_A	$6·022 \times 10^{23}$ mol^{-1}
Bohr magneton	μ_B	$9·274 \times 10^{-24}$ J T^{-1}
Bohr radius	a_0	$5·292 \times 10^{-11}$ m
Boltzmann constant	k	$1·381 \times 10^{-23}$ J K^{-1}
charge of a proton (charge of an electron = $-e$)	e	$1·602 \times 10^{-19}$ C
Faraday constant	F	$9·649 \times 10^4$ C mol^{-1}
gas constant	R	$8·314$ J K^{-1} mol^{-1}
nuclear magneton	μ_N	$5·051 \times 10^{-27}$ J T^{-1}
permeability of a vacuum	μ_0	$4\pi \times 10^{-7}$ H m^{-1} or N A^{-2}
permittivity of a vacuum	ε_0	$8·854 \times 10^{-12}$ F m^{-1}
Planck constant	h	$6·626 \times 10^{-34}$ J s
(Planck constant)/2π	$\hbar$	$1·055 \times 10^{-34}$ J s
rest mass of electron	m_e	$9·110 \times 10^{-31}$ kg
rest mass of proton	m_p	$1·673 \times 10^{-27}$ kg
speed of light in a vacuum	c	$2·998 \times 10^8$ m s^{-1}

$\ln 10 = 2·303$ $\ln x = 2·303 \lg x$ $\lg e = 0·4343$ $\pi = 3·142$
$R \ln 10 = 19·14$ J K^{-1} mol^{-1} $RTF^{-1} \ln 10 = 59·16$ mV at 298·2 K

Oxford Chemistry Series

General Editors
P. W. ATKINS   J. S. E. HOLKER   A. K. HOLLIDAY

Oxford Chemistry Series

1. K. A. McLauchlan: *Magnetic resonance*
2. J. Robbins: *Ions in solution (2): an introduction to electrochemistry*
3. R. J. Puddephatt: *The periodic table of the elements*
5. D. Whittaker: *Stereochemistry and mechanism*
7. G. Pass: *Ions in solution (3): inorganic properties*
10. J. Wormald: *Diffraction methods*
13. A. Earnshaw and T. J. Harrington: *The chemistry of the transition elements*
14. W. J. Albery: *Electrode kinetics*
18. G. C. Bond: *Heterogeneous catalysis: principles and applications*
19. R. P. H. Gasser and W. G. Richards: *Entropy and energy levels*
20. D. J. Spedding: *Air pollution*
21. P. W. Atkins: *Quanta:* a handbook of concepts
22. M. J. Pilling: *Reaction kinetics*
23. J. N. Bradley: *Fast reactions*
25. C. F. Bell: *Principles and applications of metal chelation*
26. D. A. Phipps: *Metals and metabolism*
27. J. Mann: *Secondary metabolism*
29. J. Staunton: *Primary metabolism: a mechanistic approach*
30. C. A. Coulson: *The shape and structure of molecules* (2nd edn; revised by R. McWeeny)
31. E. B. Smith: *Basic chemical thermodynamics* (3rd edn)

C. A. COULSON, F.R.S.

The shape and structure of molecules

Second edition

REVISED BY
ROY McWEENY

Clarendon Press · Oxford · 1982

Oxford University Press, Walton Street, Oxford OX2 6DP

London Glasgow New York Toronto
Delhi Bombay Calcutta Madras Karachi
Kuala Lumpur Singapore Hong Kong Tokyo
Nairobi Dar es Salaam Cape Town
Melbourne Wellington

and associate companies in
Beirut Berlin Ibadan Mexico City

Published in the United States by
Oxford University Press, New York

British Library Cataloguing in Publication Data

Coulson, C. A.
    The shape and structure of molecules.
    —2nd ed.
    1. Chemical bonds
    I. Title   II. McWeeny, Roy
    541.2'24   QD461

    ISBN 0–19–855517–2
    ISBN 0–19–855518–0 Pbk

Typeset by Cotswold Typesetting Ltd., Gloucester

Printed in Great Britain by J. W. Arrowsmith Ltd Bristol

Editor's foreword

THE explanation of the way in which atoms combine together to form molecules is one of the major triumphs of the application of quantum-mechanical ideas to chemistry, for it is possible to understand not only why atoms combine but also why the resulting molecules have their characteristic geometry. The quantum mechanics underlying the description can be complicated, but it is possible to extract the simple qualitative ideas that enable one to appreciate the factors responsible for determining the shape of a molecule. These ideas are set out in this book.

The study of the shape and structure of molecules is one of the primary meeting grounds of experiment and theory. A significant part of theoretical chemistry is concerned with the calculation of molecular structure, and experimental techniques, often of considerable sophistication, are used to provide experimental evidence. The study of molecules by *Diffraction methods* is described in a book in this series: some structural information can be obtained by *Magnetic resonance*, and information of considerable detail can be obtained by other spectroscopic techniques, as will be described in forthcoming volumes. The atomic properties that determine molecular structure are reviewed in *The periodic table of the elements*, and the implications of molecular shape for organic reactions are described in the volume on *Stereochemistry*.

P.W.A.

Preface to second edition

THE first edition of this book made frequent reference to the author's more comprehensive monograph *Valence*. Some revision was therefore called for when *Valence* ran into a third edition under the new title *Coulson's Valence*. At the same time, preparation of a second edition of the present book afforded an opportunity of responding to criticisms of the first. There were, it is true, one or two conspicuous omissions; there was no hint of the great advances in our understanding of transition-metal compounds, which now form such a large part of inorganic chemistry; nor was there any reference to the status of current theoretical methods of calculating electronic structures from first principles, in spite of magnificent progress during the previous fifteen years.

I have tried to remedy such omissions, by adding new sections, without making any heavier demands on the reader. Indeed, it was pointed out to me by some critics that certain parts of the first edition were already somewhat terse and that even the best of students would have a hard time grasping difficult concepts so briefly presented. I have therefore rewritten a few sections and passages in a rather more leisurely way, notably those dealing with orbitals, wave-mechanical principles, spin, and similar topics.

The remainder of the book is more or less unchanged. I have tried to preserve everywhere the spirit of the first edition and its high standards of clarity; and I hope the book will continue to provide, in compact form, a first glimpse of the whole field of chemical valence.

Sheffield   R.McW.
October 1981

Preface to the first edition

No one really understands the behaviour of a molecule until he knows its structure that is to say: its size, and shape, and the nature of its bonds. In recent years superb experimental developments have provided us with an immense amount of information about these very matters. It is the role of theory to make a pattern out of this information, and provide an insight into the principles of molecular architecture. This little book sets out to explain these principles, and to show show naturally much of the recent knowledge of valence angles and valence numbers follows from them. Detailed calculations would involve us in heavy computational mathematics, but fortunately extremely little mathematics is needed to understand the arguments in this book. For these arguments can be set out in pictorial form, and for that reason they are easily accepted by an interested person who wants to enjoy chemistry. This book describes the necessary wave mechanics, and then goes on to discuss, first diatomic molecules, and then polyatomic molecules. Chapter 4 reviews the concept of valence by considering the behaviour to be expected for each Group of the Periodic Table. Finally in Chapter 5 we outline those extensions of the original concept of a bond which have now proved to be necessary. It is always exciting to see order coming out of a mass of facts. The Periodic Table was itself a supreme example of this order. Its successor—chemical valence—is as wide in its scope and (dare one say it?) hardly less exciting.

July 1972   C. A. COULSON

Acknowledgements

THANKS are due to the publishers of the following journals for permission to reproduce material.

Endeavour (Figs. 1, 2), *Proc. Camb. phil. Soc.* (Fig. 10), and *Science* (Figs. 21, 35).

Contents

1. INTRODUCTION: THE SIZE AND SHAPE OF A MOLECULE 1

Size and shape. Atomic orbitals. Spin and the Pauli principle. The *Aufbau* procedure. Role of the energy in determining size and shape. Wavemechanical principles: the Hamiltonian. The variation method. The Ritz form of Rayleigh's principle. The present status of quantum chemistry. Semiempirical calculations.

2. DIATOMIC MOLECULES 19

United- and separated-atom viewpoints. The Heitler–London wavefunction for H_2. Possible refinements of the Heitler–London wavefunction. The spin factor. Homonuclear diatomics. Principle of maximum overlapping. Heteronuclear diatomics. Molecular-orbital method: H_2. Homonuclear diatomic molecules. Ionization potentials. Heteronuclear diatomic molecules. Charge distribution in molecular hydrogen.

3. POLYATOMIC MOLECULES 44

Bond properties. Hybridization. Methane. The valence state. Double and triple bonds in carbon compounds. Bent bonds: strain. Advantages and disadvantages of hybridization. Different types of hybrid.

4. VALENCE RULES 59

Valence rules—preliminary statement. Group I: the alkali atoms. Group VII atoms: the halogens. Group VI atoms. Group V atoms. Group II atoms. Group III atoms. Group IV atoms. Group VIII atoms: the rare gases.

5. DELOCALIZED BONDS 72

Many-centre bonding. Diborane. Walsh diagrams. Benzene and the world of aromatic molecules. Transition-metal compounds.

FURTHER READING 88

INDEX 91

1. Introduction: the size and shape of a molecule

Size and shape

IN 1858 A. S. Couper (1831–1892) introduced the symbol which is now universally used for a single bond (e.g. H—Cl, though he sometimes used a dotted line as in H...Cl). Three years later another Scottish chemist Crum Brown (1838–1922) used the double line in his formula for ethylene ($CH_2=CH_2$); and at about the same time Erlenmeyer used the triple line for acetylene ($HC\equiv CH$). Since then a large part of chemical effort has been put into understanding what these symbols imply. For as soon as we write a formula such as H—O—H for water we are bound to ask what geometrical relationships are involved in the two adjacent straight lines. Couper's famous 1858 paper in *The Edinburgh New Philosophical Journal* was in fact the first representation in print of what today we should call a 'structural' formula.†His work stimulated the Russian A. M. Butlerov (1828–1886) to give a lecture in 1861 entitled 'The chemical structure of compounds' and so introduced the word 'structure' to cover the ideas of size and shape. In 1865 models similar to our billiard-balls-and-springs were used by Hofmann in a lecture at the Royal Institution. Butlerov's book on organic chemistry published in 1864 and Kekulé's in 1865 completed the work. Structural studies were to become an increasingly important part of almost every branch of chemistry.

Structural chemistry is concerned with valence—why atoms combine in definite ratios, and how this is related to the bond directions and bond lengths. In its simplest terms this is the question: what do we mean by a chemical bond? We shall, therefore, be concerned with such matters as size and shape. Now when we talk about size and shape we have to be careful. For we may be referring to the positions of the nuclei, or we may be referring to the distribution of electronic charge. Nuclei are always vibrating, since even at the agsolute zero of temperature they can never avoid their zero-point motion. However, such vibrational amplitudes seldom exceed about 10 pm (ten picometres or 10^{-11} m), whereas most bond lengths are in the range 100 to 300 pm so that for nearly all structural purposes we can think of the nuclei in terms of their mean positions. The electrons, on the other hand, are best thought of as a charge-cloud, whose density varies from place to place, and is determined from the wavefunction (see later). The shape of a bond, therefore, is really the shape of this charge-cloud—more precisely, that part of the total charge-cloud which we can associate with the bond.

† By accident Couper's paper appeared in print in a French version a few weeks after a somewhat similar paper by F. A. Kekulé (1829–1896). Both writers discussed the quadrivalence of carbon, but only Couper used the line symbol for a bond.

Experimentally the positions of the nuclei are found either from a study of vibration–rotation spectra, which provides three principal moments of inertia, or by use of neutron diffraction. In this latter technique a homogeneous beam of neutrons is scattered by the molecule: in this scattering process the nuclei predominate. The details of the electron-cloud distribution are obtained from X-ray scattering, for X-rays are preferentially scattered by electrons; and from measurements of the intensity at different angles of scatter it is possible to infer the charge density of the scatterer. X-ray scattering is usually limited to regular solids, i.e. crystals. In the gas phase electron diffraction is used instead. In particular circumstances, electron-spin-resonance techniques, or techniques making use of the coupling between two nuclear spins (e.g. of two hydrogen atoms) which depends on the internuclear distance, can provide additional information. In the rest of this book we shall freely make use of experimental results obtained by one or other of these methods. An excellent summary of structural determination is provided by two Chemical Society publications (Sutton 1958, 1965).†

The most familiar way of representing the details of the charge-cloud is by means of a contour diagram. Figure 1 shows such a diagram for anthracene. Each contour joins points in the molecular plane at which the total charge density has the same value. There are obvious peaks at the positions of the fourteen carbon nuclei, from which the internuclear distance (i.e. bond lengths) can be inferred. It will be noticed that on this diagram the positions of the hydrogen atoms are far from precise. This is because a hydrogen atom provides only one electron compared with six from a carbon atom: so the intensity of X-rays scattered from that part of the molecule is small, leading to considerable uncertainty in position. However, if we use neutrons, then as Fig. 2 shows for the case of benzene, the situation is quite different, and the

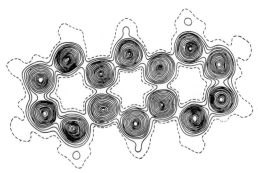

Fɪɢ. 1. Electron-density contours for anthracene determined by X-ray methods. (*Endeavour* **25**, 129 (1969) courtesy of G. E. Bacon.)

† The uses of neutron, X-ray, and electron diffraction are discussed by J. Wormald in *Diffraction methods* (OCS 10). Electron spin resonance and nuclear magnetic resonance are described by K. A. McLauchlan in *Magnetic resonance* (OCS 1).

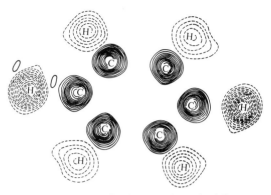

Fig. 2. Nuclear-density contours for benzene determined by neutron methods. (*Endeavour* **25**, 129 (1969) courtesy of G. E. Bacon.) The greater density for the top and bottom hydrogens results from the fact that, in this projection, protons from two molecules are almost superposed at these two positions.

positions of the hydrogen nuclei now show up clearly. A fully study of size and shape is therefore greatly helped if we can measure both types of diffraction for our chosen molecule.

Atomic orbitals†

Since molecules are made from atoms it is important first to remind ourselves of some of the known facts about them. Moreover, on account of the single nucleus which dominates the atom and allows all sorts of simplifications not possible for a molecule, we can make very good theoretical calculations for atoms. The simplest, and at the same time the most pictorial, of these is the orbital model. In this model we describe each electron by an orbital, or 'personal wavefunction'; the magnitude of this wavefunction varies from point to point in space, and if x, y, z are the Cartesian coordinates of a point we write the wavefunction as $\phi(x, y, z)$—a function of position. When the state of an electron is described by ϕ the position of the electron is uncertain but $|\phi(x, y, z)|^2$ gives the *probability per unit volume* of finding it at point x, y, z. In other words $|\phi|^2 \, d\tau$, where $d\tau$ denotes a small element of volume, gives the probability of the electron being in $d\tau$; this may be visualized as the fractional amount of time it spends in $d\tau$ or the amount of *charge* (in units of an electron) associated with $d\tau$. For many purposes an electron in orbital ϕ may thus be pictured as a 'charge cloud', as if it were

† More details may be found in *Coulson's Valence*, abbreviated in subsequent references to *CV*. (See Further Reading p. 88.)

smeared out in space with a density ϕ^2. In the orbital model of a many-electron atom the wavefunction is approximated by a product of orbital factors, one for each electron, and† the total charge cloud is obtained simply by summing the individual orbital contributions i.e. by superimposing the *orbital* charge clouds. This is true for molecules, as well as for atoms, and is what gives the orbital model its great conceptual value.

For atoms, the atomic orbitals (AOs) have symmetry properties similar to those which describe the possible states (ground state and excited states) of the single electron in a hydrogen atom; and these properties are dictated by *quantum numbers*. The shapes of the AOs are determined by an 'angular momentum quantum number' $l = 0, 1, 2, \ldots$, successive values yielding orbitals of s, p, d, ... type (a designation arising historically from the names of spectroscopic series). If we use $r(= \sqrt{(x^2 + y^2 + z^2)})$ to denote the distance of the electron from the nucleus, the s-orbitals are all functions of r alone and therefore show no dependence on angle—they are spherically symmetrical (Fig. 3 (a, b, c)). The s-functions fall away exponentially as r increases and the charge cloud therefore rapidly becomes thinner at large distances from the nucleus. The different s-orbitals are labelled by a principal quantum number $n = 1, 2, 3, \ldots$ and designated 1s, 2s, 3s, ... In the 1s orbital the charge density falls away smoothly, going out from the nucleus, but in the 2s, 3s, ... states, which correspond to successively higher values of the energy, the density falls to zero at one, two, or more, intermediate values of r; these points correspond to *nodes* in the wavefunction, separating regions in which ϕ has positive and negative values. As the number of oscillations increases, the orbital becomes larger‡ (i.e. more diffuse) and its energy increases.

The p-functions ($l = 1$) occur *in sets of three* (Fig. 3 (d, e, d)), each set describing three alternative states in which the electron has exactly the same energy; three such p-orbitals, which we denote by p_x, p_y, p_z are said to be *triply degenerate*. Again using x, y, z for the Cartesian coordinates of the electron, and r for its distance from the nucleus, then a typical set of p-orbitals is

$$p_x = xf(r), \quad p_y = yf(r), \quad p_z = zf(r), \tag{1}$$

where $f(r)$ is a spherically symmetrical function. As in the case of s-orbitals there are sets of p-orbitals corresponding to different choices of principal quantum number ($n = 2, 3, \ldots$); these we designate by 2p, 3p, ... The values $n = 1, 2, 3, \ldots$ determine the K, L, M, ... 'shells' of the atom; for hydrogen,

---

† Provided the orbitals are correctly chosen, a point to which we return later.

‡ The 'size' of an orbital for an electron moving around a central charge Ze (e.g. $Z = 1$ for H, $Z = 2$ for He$^+$), is of considerable importance. The radial factor, denoted by $f(r)$ in all the AOs we discuss, behaves for larger r like $r^{n-1} \exp(-Zr/na_0)$, where n is the principal quantum number and a_0 is the Bohr radius (see back end-paper). From this it follows that the outermost peak in the charge density (ϕ^2) occurs at a distance $r_{max} \simeq n^2 a_0/Z$. This quantity gives a rough measure of size; clearly as the central charge increases the charge cloud is drawn tightly in towards the nucleus, while 1s, 2s, 3s, ... AOs (or 2p, 3p, ... etc.) get larger roughly in proportion to n^2.

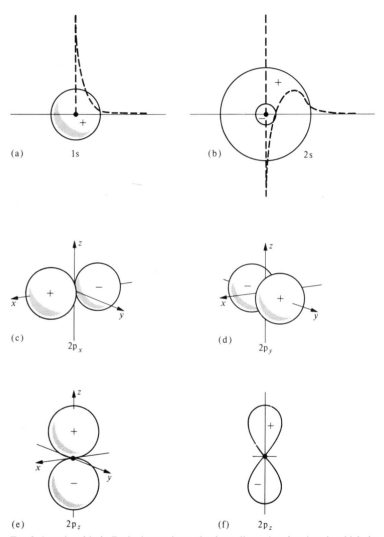

FIG. 3. Atomic orbitals. Each picture shows the three-dimensional regions in which the wavefunction ϕ has a substantial value, the sign of ϕ being indicated by + or −. Outside these regions ϕ falls exponentially to zero. In the first two pictures, ϕ is plotted (broken lines) against distance of the electron from the nucleus. The last picture (f) shows an even more schematic representation of a 2p AO.

all the AOs in a given shell have identical energy and are roughly similar in size. It should be noted that there is no 1p AO; there is a rule that n must always be greater than l and consequently the first p-orbitals ($l = 1$) are in the second (or L) shell with $n = 2$.

The d-orbitals ($l = 2$) which are so important in transition-metal chemistry, occur in sets of five (Fig. 4). The most usual choice of five independent functions is

$$d_{xy} = xyf(r), \ d_{yz} = yzf(r), \ d_{zx} = zxf(r),$$
$$d_{x^2-y^2} = \tfrac{1}{2}(x^2 - y^2)f(r), \ d_{z^2} = \sqrt{\tfrac{1}{12}}\,(3z^2 - r^2)f(r) \tag{2}$$

where $f(r)$ is the same function of r in all cases, though differing in form from that for the other types of orbital (e.g. the p-functions, above) and again depending on the principal quantum number. The first d-orbitals ($l = 2$) appear in the third (or M) shell. In the lanthanides and actinides, f-orbitals (which first appear in the shell with $n = 4$) are also important, but we shall not need to use them in this book.

For a hydrogen atom the AO energies depend only on the principal quantum number and thus fall into groups; in ascending energy order they are

$$1s < (2s, 2p) < (3s, 3p, 3d) < (4s, 4p, 4d, 4f) < \ldots.$$

For many-electron atoms we find this order is disturbed.

One property of the above atomic orbitals, which are particular solutions of Schrödinger's famous wave equation, is of special importance; if we

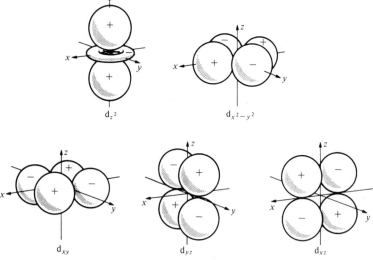

Fig. 4. Family of five d-type AOs.

multiply together the values of two different AOs centred on the same atom, ϕ_1 and ϕ_2, within every small volume element $d\tau$, and then sum over the whole of space to get the integral $\int \phi_1 \phi_2 \, d\tau$, *the result will be zero*. This property is called 'orthogonality' and the same property may be assumed for all solutions of the Schrödinger equation, whether they refer to an atom or a molecule. Orbitals must be orthogonal if we are to talk meaningfully about them separately. If two orbitals are *not* orthogonal we may say that either one of them 'contains' part of the other or that they have a mutual 'overlap'; and in that case it turns out that the total charge density cannot be represented as a sum of contributions ϕ^2 from the individual orbitals. The orthogonality property is thus a cornerstone of the orbital model of atoms and molecules, on which all simple pictorial descriptions ultimately depend. Whenever there is any ambiguity in the choice of a set of orbitals we should try to make them orthogonal; all the above orbitals, some of which are shown in Fig. 3, satisfy this requirement. Orbitals are often orthogonal because of a difference of symmetry which ensures that a $\phi_1 \phi_2$ value at one point in space is always exactly cancelled by an equal and opposite value at another point; this is the case, for example, for 2s and $2p_x$ (Fig. 3)†. They may also be orthogonal when they lie mainly in different regions of space (e.g. a 'big' orbital and a 'small' orbital) and have a region of overlap in which positive and negative contributions of $\int \phi_1 \phi_2 \, d\tau$ balance out exactly, even in the absence of any difference of symmetry; this is the case for 1s and 2s. We shall find many other examples of orthogonality. In addition to their orthogonality it should be noted that orbitals are usually *normalized* so that $\int \phi^2 \, d\tau = 1$; since ϕ^2 may be visualized as a charge density associated with an electron in ϕ, this simply means that the total 'smeared out' charge when integrated over all space must yield just one electron.

Now that we know how to describe the individual electrons in an atom, we are almost ready to use the *Aufbau* (German for 'building up') approach to predict the electronic structure of atoms in the periodic table. First, however, we must ask why atoms in their lowest-energy, or ground states do not all have essentially the same electronic structure—with all electrons in the same lowest-energy orbital (1s). There must be some principle which prevents our allocating all electrons to the same orbital and thus accounts for the rich variety of structures actually found; this principle, first adopted in the days before wave mechanics, is the Pauli exclusion principle.

Spin and the Pauli principle

Experiment shows that even the ground state of the hydrogen atom, with one electron is a 1s orbital, is a *doublet* state i.e. the electron may be in either one of two possible states. These states are distinguished by the *spin* of the

---

† Imagine 2s and 2p superimposed on the same centre and compare the values of their product $(\phi_1 \phi_2)$ at any point and its mirror image across the yz-plane.

electron; if a magnetic field is applied the spin (which has an associated magnetic dipole) may line up parallel or anti-parallel to the field ($\uparrow$ or $\downarrow$). The two spin states are recognized by attaching a 'spin factor', α or β, to the orbital; the two possible states for an electron in orbital ϕ are then described by $\phi\alpha$ ('spin up') and $\phi\beta$ ('spin down'). In the presence of an applied magnetic field the two states differ very slightly in energy, as is revealed by electron spin resonance (e.s.r.) experiments,† but otherwise they are degenerate. When spin is taken into account, every orbital gives rise to two *spin-orbitals* and the available electronic states for an electron moving around a nucleus thus become 1sα, 1sβ, 2sα, 2sβ, 2p$_x\alpha$, 2p$_x\beta$, ... Pauli's famous exclusion principle then takes a very simple form: *in the orbital description of an atom no two electrons can occupy the same spin-orbital.* Alternatively, two electrons can be put into the same orbital (such as 2s) but they must then differ in spin (2sα and 2sβ); colloquially, an orbital can hold *two* electrons but not more than two.

In its modern form, Pauli's principle is usually stated in a more general way; the many-electron wavefunction, with spin included, must be antisymmetrical under interchange of any two electrons. To see what this means let us put two electrons into a 1s orbital. If we use 1 and 2 to indicate the coordinates of electron 1 and electron 2, respectively, a suitable 2-electron wavefunction (in the absence of spin) would be the product 1s(1)1s(2); on including spin, we should have to use a spin-orbital product such as 1sα(1)1sβ(2) or 1sα(2)1sβ(2). The modern form of Pauli's principle tells us that neither of these products is in itself an acceptable wavefunction; but from the two, which differ only by interchange of electrons 1 and 2, we can construct

$$\Phi(1, 2) = [1s\alpha(1)1s\beta(2) - 1s\alpha(2)1s\beta(1)]$$

which changes sign on switching 1 and 2. An 'anti-symmetrized product' can always be constructed from a product in which the electrons are assigned to different spin-orbitals. If, however, we tried to put two electrons into the *same* spin-orbital, 1sα say, any attempt to antisymmetrize would destroy the wavefunction; for we should obtain

$$\Phi(1, 2) = [1s\alpha(1)1s\alpha(2) - 1s\alpha(2)1s\alpha(1)] = 0$$

In other words we cannot find acceptable wavefunctions which would describe two (or more) electrons in the same spin-orbital; and if two are in the *same* orbital then they must have opposite spins. Pauli's pre-wave-mechanical exclusion principle is thus a simple consequence of the antisymmetry principle.

It might be said that the whole of chemistry rests upon the fact that electrons obey the Pauli principle; for without it, as we have remarked, all the electrons of any atom could be assigned to the same 1s orbital—there would be no variety, no periodicity of properties, no periodic table! We must now ask what actually happens when we systematically 'build up' an atom by starting with a bare nucleus of charge Ze and adding Z electrons.

The *Aufbau* procedure

With each AO we have an associated orbital energy—the energy of an

---

† See, for example, K. A. McLauchlan in *Magnetic resonance* (OCS 1).

electron described by (or 'occupying') that orbital. The conventional zero of the energy corresponds to an electron at rest at infinity i.e. taken away from the atom, leaving an ion. The orbital energies are thus negative, as indicated in Fig. 5 which shows the usual sequence of levels in a typical atom, namely

$$\varepsilon_{1s} < \varepsilon_{2s} < \varepsilon_{2p} < \varepsilon_{3s} < \varepsilon_{3p} < \varepsilon_{4s} \approx \varepsilon_{3d} < \varepsilon_{4p} \ldots.$$

And the distance of any level below the 'ionization limit' $\varepsilon = 0$ (i.e. the magnitude of an orbital energy) represents the minimum amount of energy needed to ionize the atom by taking an electron out of the corresponding orbital—an orbital ionization potential. We shall see later (p. 35) how such ionization potentials can be obtained experimentally.

The 3d orbital energy sometimes lies below the 4s energy, especially towards the end of the first transition-metal series: so also the 4d lies below the 5s towards the end of the second transition series. The *one*-electron atoms of H, He^+, Li^{++} ... are exceptions to the sequence indicated above since for these, and only for these, the 2s and 2p AOs are degenerate, as are the AOs of any given principal quantum number.

To obtain the *electron* configuration of an atom in its ground state we now simply fill up the orbitals, feeding in the electrons two at a time (to observe the Pauli principle) in the ascending energy order shown in Fig. 5, and continuing until all electrons have been allocated. This is the *Aufbau* procedure. A few examples are given below for the ground states of the lighter atoms, superscript 2 indicating an orbital occupied by two electrons with 'paired' spins ($\uparrow\downarrow$):

H (1s)   C $(1s)^2(2s)^2(2p_x)(2p_y)$
He(1s)^2 N $(1s)^2(2s)^2(2p_x)(2p_y)(2p_z)$
Li $(1s)^2(2s)$ O $(1s)^2(2s)^2(2p_x)(2p_y)(2p_z)^2$
Be $(1s)^2(2s)^2$ F $(1s)^2(2s)^2(2p_x)(2p_y)^2(2p_z)^2$
B $(1s)^2(2s)^2(2p_z)$ Ne $(1s)^2(2s)^2(2p_x)^2(2p_y)^2(2p_z)^2$
Na(K)(L)(3s)   Mg(K) (L) $(3s)^2$

The first five elements present no difficulties, except that in boron the solitary unpaired electron may be in any one of the three equivalent 2p AOs. It is customary in such cases to call it the $2p_z$ AO. With carbon, where there are two electrons to be located in one or more of the degenerate 2p orbitals, we cannot at this stage say whether to put both in the same orbital, with opposed spins, or in different orbitals, with opposed or with parallel spins. This matter is resolved by Hund's rules, which state:

(i) with degenerate, or nearly degenerate, orbitals, electrons prefer to be located in different AOs rather than the same AO, since this keeps them rather further apart, and reduces the repulsion between them;

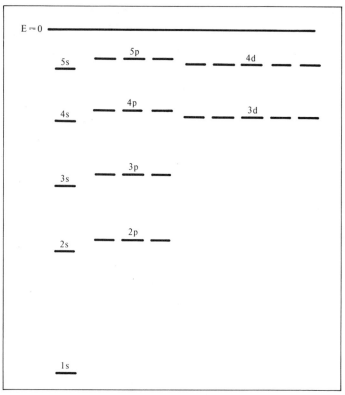

Fig. 5. Orbital-energy diagram for atoms. Each line (or orbital) will hold up to two electrons. The vertical scale is not realistic; otherwise the 1s level would lie below the bottom of the page.

(ii) with degenerate, or nearly degenerate, orbitals, electrons in different AOs prefer to have parallel spins, since this leads to a favourable exchange interaction.†

Using these rules we see that the lowest configurations for C, N, O, and F are simply determined. Neon completes the L-shell, and sodium, with an electronic configuration similar to that of lithium, starts the M-shell. All this will be very familiar to most readers, but we shall find that much the same kind of discussion also applies to molecules.

---

† The name arises from the fact that an antisymmetric many-electron wavefunction changes sign under exchange of two electrons (p. 8); 'exchange interaction' merely describes an energy term associated with the antisymmetry requirement − and not in any sense a real physical effect like an electrostatic interaction.

Role of the energy in determining size and shape

It is found experimentally that the water molecule H_2O is triangular in shape, with a valence angle of $104\frac{1}{2}°$, but that CO_2 is linear. The simplest way of explaining this is to say that if we choose different positions for the three atoms, and then calculate the energy of the electrons, we find that in H_2O the total energy is least for a triangular shape, whereas for CO_2 it is least for a linear shape. So questions of molecular shape are really questions of molecular energy. So also is the matter of bond length. The total energy in H_2O is lowest if, in addition to the valence angle being $104\frac{1}{2}°$, the two bond lengths are equal and have a value 96 pm. This means that the overall size and shape of the molecule are governed by the total energy. In later chapters we shall therefore have to study this carefully, since conventional valence rules are simply rules derived from the way in which the molecular energy varies with the number and positions of the various atoms.

The total energy of a molecule is usually estimated by solving the Schrödinger wave equation (almost always approximately) to determine a wavefunction and energy for the electrons, the nuclei being assumed fixed in their equilibrium positions, and then adding on the electrostatic repulsion energy of the nuclei. But why do we consider explicitly only the electrons? The answer to this question is provided by an important principle due to Born and Oppenheimer. They wrote down the full wave equation, to determine the possible states of the whole system of particles, and noted that terms involving nuclear motion resembled those for the electronic motion but were multiplied by a factor m_e/m_N, a ratio of electronic and nuclear masses. Now even for the single proton of a hydrogen atom this ratio is no more than $\frac{1}{1840}$; and it is therefore not surprising that Born and Oppenheimer, by expanding the equation in powers of the mass ratios, were able to show that the effect of nuclear motion could be taken into account by adding small nuclear kinetic energy terms to the energy in fixed-nucleus approximation, any remaining corrections being exceedingly small. In physical terms, we might say that the heavy nuclei move so slowly compared with the electrons that, to an electron, they appear to be 'frozen' in their instantaneous positions. The energy of the electrons may thus be calculated for any fixed configuration of the nuclei and the remaining energy will be almost entirely electrostatic repulsion energy of the nuclei, the nuclear motion effects being added subsequently. In this way, for example, we can quite properly talk about the electronic energy† of a diatomic molecule corresponding to any internuclear distance R, obtaining a familiar 'potential energy curve' of the kind drawn in Fig. 6. If we want to explain why two hydrogen atoms form a stable diatomic molecule H_2, we must be able to calculate a curve similar to that of the figure, and use it to infer the equilibrium bond length R_e and dissociation energy D_e.

---

† To be more precise, such curves usually show the *total* energy in fixed-nucleus approximation, the nuclear repulsion energy being included, by convention, in the 'electronic' energy.

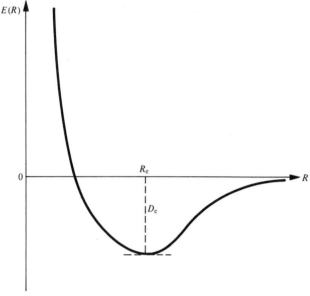

FIG. 6. Potential-energy curve for a diatomic molecule. R_e is the equilibrium bond length, D_e is the dissociation energy, no account being taken of the zero-point vibrational motion.

With polyatomic molecules there are more internal coordinates than just the bond length, as in H_2. For example, in a hypothetical complex H_3 of three hydrogen atoms, we should have an energy surface in which the molecular energy depended on two bond lengths and one valence angle (or, if we preferred, three bond lengths: but for most purposes it is better to keep to valence angles where possible). Then, in order to show that no stable H_3 molecule could form, all that would be necessary would be to show that this energy surface had no minimum such as that in Fig. 6. Assuming that our calculations were sufficiently good, this would guarantee that no stable state of H_3 could exist.

In the same sort of way we have to show that methane CH_4 is a tetrahedral molecule by proving that the lowest value of the total molecular energy occurs when the four hydrogen atoms are tetrahedrally oriented around the central carbon atom.

Thus the importance of the Born–Oppenheimer approximation is very considerable. It *is* an approximation. But fortunately the corrections needed in order to go beyond it are so very small that we can safely neglect them in any theory of valence.†

† For H_2^+ the error is only 720 J mol^{-1} (0.0075 eV) whereas the dissociation energy is 266 kJ mol^{-1} (2.77 eV); the ratio of these numbers is about 370:1.

Wave-mechanical principles: the Hamiltonian

The rest of this chapter will be devoted to a brief resumé of those parts of wave mechanics that will be needed later. Fortunately this is not very extensive. It is not difficult to write down the precise form of the Schrödinger wave equation for any given problem. In an extremely important and elegant way we write it

$$H\psi = E\psi \tag{3}$$

where the Hamiltonian H is merely the wave-mechanical transcription of the kinetic plus potential energies of all parts of the system. The potential energy presents no problem since, in the absence of any external electric or magnetic fields, it is simply the sum of all Coulomb interactions between each pair of positive and negative charges. So we obtain a sum of terms such as $e_i e_j/4\pi\varepsilon_0 r_{ij}$, where r_{ij} is the distance between point charges e_i and e_j. The kinetic energy is the sum of the kinetic energy of each particle. A mass m_i at the point (x_i, y_i, z_i) then makes a contribution $-(h^2/8\pi^2 m_i)\nabla_i^2$, where. $\nabla_i^2 \equiv \partial^2/\partial x_i^2 + \partial^2/\partial y_i^2 + \partial^2/\partial z_i^2$. It is in this way that H is obtained, however many particles there may be. Fortunately, however, as we have seen, the Born–Oppenheimer approximation enables us to forget the nuclear kinetic-energy terms, proceeding as if the nuclei were fixed in their equilibrium positions. In future, therefore, we shall use H to imply only that part of the full Hamiltonian that remains when these terms are omitted.

The variation method

It is clear that the wave equation, while easy to write down, is impossibly difficult to solve exactly. Indeed, only for the case of one electron and at most two nuclei can this be done. This forces us to use approximations. We have to remember that we are not just trying to solve the appropriate partial differential equation, we are trying to find those values of the energy E† for which acceptable solutions of the equation exist. For this purpose some form of the variation method, originally developed in the 1880s by Lord Rayleigh, and therefore often referred to as Rayleigh's principle, is almost always used.

It follows, after multiplying both sides of the wave equation $H\psi = E\psi$ by ψ^* (ψ^* is complex conjugate to ψ: if ψ is real, as is usually the case in this book, $\psi^* \equiv \psi$) and integrating, that

$$E = \int \psi^* H\psi \, d\tau \Big/ \int \psi^* \psi \, d\tau, \tag{4}$$

where $d\tau$ implies integration over all the coordinates involved in the problem, including spins. Eqn (4) is not much use as it stands since in order to use it to

---

† These are called the *eigenvalues* of H: the corresponding ψ are the *eigenfunctions*.

obtain E we should need to know ψ: and if we once knew ψ we could as soon use the equation $H\psi = E\psi$ to determine E. But Lord Rayleigh showed that even if we do not know the true ψ we can still make progress. Let us introduce a 'trial function' ψ, which is most unlikely to be the true eigenfunction; and let us use eqn (4) to define the *Rayleigh ratio* $\mathscr{E}\{\psi\}$ by the relation:

$$\mathscr{E}\{\psi\} = \int \psi^* H\psi \, d\tau \Big/ \int \psi^* \psi \, d\tau. \tag{5}$$

We have written the left hand side $\mathscr{E}\{\psi\}$ to show that the value of the Rayleigh ratio, which certainly has the dimensions of energy, will depend on what *function* ψ we have chosen. The important property of $\mathscr{E}\{\psi\}$ is that for the ground state of the system† $\mathscr{E}\{\psi\} \geqslant E_0$, where E_0 is the true ground-state energy. By testing various ψ and then selecting that one which gives the lowest value of $\mathscr{E}\{\psi\}$ we get the best approximation to the true energy and true wave function. If we are in any doubt as to whether we have done the best that we can, we can always try one or two new trial functions. If they give a lower value of $\mathscr{E}\{\psi\}$ they are improvements on what we have found before: otherwise in general they are not.

It is very time-consuming to try many different ψs. So we usually build into our trial function a certain amount of flexibility. This is generally in the form of one or more parameters $c_1, c_2, \ldots c_n$. In this way $\mathscr{E}\{\psi\}$ becomes some function of these parameters. Their values are then chosen so that $\mathscr{E}$ is stationary (minimum). We have to find the values of $c_1, c_2 \ldots$ *such that* $\partial\mathscr{E}/\partial c_1 = 0 = \partial\mathscr{E}/\partial c_2 = \ldots$. It can be shown that adding more parameters improves the result, so that the choice lies between more numerical work leading to better E and ψ, or less numerical work and a poorer E and ψ. Fortunately, as we shall see in later chapters, for the purposes of understanding molecular shape and size, only two or three such variable parameters are usually necessary.

An example will show how this works. Consider the ground state of the helium atom. With the notation of Fig. 7 the Hamiltonian (see p. 13) is

$$H = -\frac{h^2}{8\pi^2 m}(\nabla_1^2 + \nabla_2^2) - \frac{e^2}{4\pi\varepsilon_0}\left(\frac{2}{r_1} + \frac{2}{r_2} - \frac{1}{r_{12}}\right).$$

The *Aufbau* principle of p. 9 suggests that since the orbital description is $(1s)^2$ we should try a wave function of the form

$$\psi(1, 2) = e^{-cr_1/a_0} \times e^{-cr_2/a_0}$$

The parameter c will determine the size of the atomic orbital, as we know from the discussion on p. 4. Alternatively we could interpret this by saying that each electron appears to move independently of the other electron in the field of a central charge ce instead of the actual charge Ze. We must now

† And, in fact, for the lowest state of any symmetry. This need not necessarily be the ground state of the system.

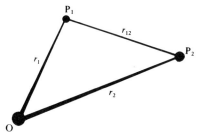

FIG. 7. Notation for the helium atom He, P_1 and P_2 are the two electrons; O is the nucleus, taken to be fixed.

calculate $\mathscr{E}\{\psi\}$. It will be simply a function of the variable parameter c. Detailed calculation† shows that

$$\mathscr{E}\{\psi\} = (c^2 - \tfrac{27}{8}c) \times e^2/4\pi\varepsilon_0 a_0.$$

This has a minimum when $d\mathscr{E}/dc = 0$, i.e. $c = 27/16$. The corresponding energy is $-2.848 \times e^2/4\pi\varepsilon_0 a_0$, whereas the experimental value is $-2.904 \times e^2/4\pi\varepsilon_0 a_0$. The agreement is really quite good. The wave function for each electron, obtained in this way, is called a *Slater-type orbital* (STO). General and widely used rules exist‡ for writing down generalizations of this type of atomic orbital for heavier atoms. The fact that the optimum value of c is $(Z - \tfrac{5}{16})$, with $Z = 2$, shows that the repulsion between the electrons effectively reduces the central field 'felt' by either; we say each electron 'screens' the other from the full attraction of the nucleus and the $\tfrac{5}{16}$ is called a 'screening constant'. The parameter c is sometimes called an 'effective Z' and is denoted by Z_e.

The Ritz form of Rayleigh's principle

In 1908 a young mathematician Ritz showed that Rayleigh's method could be put into an exceedingly simple form if the variable parameters in the trial function occurred linearly, as multiples of a set of chosen functions, or basis set. Thus we write

$$\psi = c_1\phi_1 + c_2\phi_2 + \ldots + c_n\phi_n, \tag{6}$$

where the basis set $\phi_1 \ldots \phi_n$ is chosen at the start, and only the parameters $c_1 \ldots c_n$ are varied. The advantage of this type of expansion is that it leads§ to a very simple set of linear equations involving the coefficients. By eliminating

† See, e.g. Eyring, H., Walter, J., *and* Kimball, G. E. (1944), *Quantum chemistry*, Wiley, Chapter 7.
‡ See e.g. *CV*, pp. 42–5.
§ *CV*, Chapter 3, Section 8.

the c's we obtain a secular determinant whose roots are the desired approximations to the lowest n energies. The technique lends itself so well to machine computation that it is almost universally employed in modern calculations. Moreover it is well-suited to problems in chemistry since we are at liberty to choose whatever basis set ϕ_r that we wish. In any given situation we shall naturally make as much use as we can of the intuitive ideas of chemistry, and select as basis sets those functions which embody characteristics that we believe should be found in the real situation. In later chapters we shall repeatedly make use of this idea. Thus the development of covalent--ionic resonance in Chapter 2, and the expansion of molecular orbitals in terms of atomic orbitals in Chapters 3 and 5, are just two examples of this use of the Rayleigh–Ritz method.

Present status of quantum chemistry†

From what has been said already it must be clear that any theoretical discussion of the electronic structures of atoms and molecules will rest very heavily on our ability to solve, at some level of approximation, the Schrödinger equation. During the last fifty years a whole new branch of chemistry—quantum chemistry—has evolved as a result of our efforts to solve the wave equation. The theory of valence is only one part of this subject, for the laws of quantum mechanics (of which wave mechanics is merely one particular formulation) allow us, in principle, to predict not only the electronic structure and the geometry of a molecule but indeed all its properties. With the advent of the computer, quantum chemistry has made spectacular advances. Calculations which were almost unthinkable twenty-five years ago are now almost routine and there are few areas of chemistry into which quantum theory has not penetrated.

What, then, is the status of such calculations? The answer to this question depends on their degree of refinement. At the highest level, we may attempt a highly accurate numerical solution of the Schrödinger equation; such calculations, which usually start from nothing more than a conjectured molecular geometry, are usually termed *ab initio*. At a lower level, we may set up a highly simplified 'model' of the molecule, based on empirical principles and approximations and perhaps containing adjustable parameters with values chosen by appeal to experiment; such theories are generally termed semi-empirical. Both approaches are firmly rooted in quantum mechanics and have much to contribute; which one to use depends on molecular size, the amount of computing power available, and the objectives of the investigation; but before accepting the predictions to which they lead we should be aware of their limitations—which we now briefly consider.

---

† Some of the material in this section is taken from the author's Inaugural Lecture delivered at the University of Oxford on 13 February 1973.

Ab initio calculations

The wave equation is a difficult thing to solve. Thus for water it is a partial differential equation in 39 distinct variables, even when we leave out electron spins, and we must not expect to obtain any kind of exact solution. This is hardly surprising in view of the fact that mathematicians have made very little progress with equations containing more than four independent variables.

Nevertheless, using the variation method described above, we can obtain numerical solutions of very high precision—at least for small molecules. Computers are admirable servants, and they grow steadily more rapid in their working. Since 1960 the time needed for a typical molecular calculation has dropped by a factor of 10^7 or more and, curiously, the cost per operation has also dropped. Consequently, it is now sometimes feasible to calculate molecular properties with an accuracy as great as, or even greater than, that of the best experiment.

Two examples must suffice. The observed ionization potential of the helium atom is $198\,310.82 \pm 0.15\,\text{cm}^{-1}$ whereas the computed value is $198\,310.665 \pm 0.056\,\text{cm}^{-1}$; the theoretical value thus lies within the confidence limits of the experimental value. Similarly, the best experimental bounds for the dissociation energy of the molecule H_2 are $36\,116.3\,\text{cm}^{-1}$ and $36\,118.3\,\text{cm}^{-1}$; the computed value is $36\,117.7$, again within the limits of experimental error. In this case, surprisingly, so much confidence was placed on the theoretical value that an earlier experimental result of $36\,113\,\text{cm}^{-1}$ had to be re-examined, and subsequently corrected, by the spectroscopists!

With larger molecules the accuracy attainable drops, but it is still good enough to settle many issues of great practical importance and the scope of *ab initio* calculations is constantly increasing. Thus, the methylene radical CH_2, which plays an important part in many chemical reactions, has a lifetime of less than one-millionth of a second. It is therefore out of the question to make the usual type of careful measurement of its structure. Calculation, however, shows that in the two lowest energy states, a triplet and a singlet, the CH_2 valence angles are respectively about 140° and 103°. It is remarkable that such calculations are nowadays widely accepted as a legitimate way of getting information that is experimentally inaccessible.

Semi-empirical calculations

Most of chemistry is concerned with comparatively large molecules, for which we cannot hope to get the numerical answers absolutely right. But most of chemistry is also concerned, fortunately, not with precise numerical values but rather with concepts, trends, and what have been called 'primitive patterns of understanding'. When we lower our sights to this level, asking for only enough numerical accuracy to be confident that our simplified theoretical model is substantially correct, we can often obtain a profound and deeply satisfying insight into a wide range of chemical phenomena.

Calculations of this kind are very widely used; they help to provide some theoretical justification for the more qualitative arguments used throughout this book; but they are suggestive rather than definitive, they provide the concepts and the vocabulary of quantum chemistry but do not give numerical results with any kind of absolute significance. The simple models lead more readily to a theoretical rationale of what we observe and, as long as we recognize their limitations, they provide an excellent tool for breaking new ground.

2. Diatomic molecules

United- and separated-atom viewpoints

LET us bring two atoms together from a large distance apart. If we suppose that these atoms are able to form a stable diatomic molecule we shall obtain a potential-energy (PE) curve such as that shown in Fig. 8. It will be convenient to divide this curve into three regions, of which the middle one corresponds to the true molecular situation. We shall find that as we approach this middle region from either of the two extreme regions we are led to two distinct types of approximate wave functions.

When the nuclei are far apart we speak of the separated-atom region. Here the two atoms retain many of their own characteristics, though of course they are perturbed, or polarized, by each other. The quantization in this region is essentially atomic. On the other hand, when the nuclei are close together the whole system resembles one single united atom. The quantization in this region is that of one centre. Both these extreme situations provide a model for the intermediate region of true molecular quantization. The separated-atom model leads us to the valence-bond (VB) approximation, of Heitler and London; the united-atom model leads us to the molecular-orbital (MO) approximation, due to Hund, Mulliken, and Lennard-Jones. In this chapter we shall introduce both approximations, though later we shall concentrate almost entirely on the MO approach which has both mathematical and pictorial advantages; fortunately the two approaches converge to the same

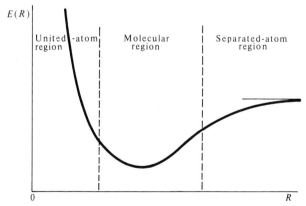

FIG. 8. The unit-atom and separated-atom regions of a molecule, and the true molecular region between them.

end point, as each is refined, and it is therefore sufficient to develop only one of them in detail. We begin with molecular hydrogen. This system is a prototype for homonuclear diatomic molecules, and of course, since there are only two electrons, it is possible to make calculations not yet possible for heavier systems.

The Heitler–London wavefunction for H_2

Let us start with the separated-atom situation of Fig. 8. If the atoms are far enough apart there is effectively no interaction between them. We may suppose that electron 1 is around nucleus A and electron 2 is around nucleus B. They must obviously have personal wavefunctions such as would be found for an isolated hydrogen atom in its ground state. If we call these two AOs ϕ_a and ϕ_b, then we should begin by expecting a combined wavefunction $\phi_a(1)\phi_b(2)$, where the symbol $\phi_a(1)$ implies that it is electron 1 that is in atomic orbital ϕ_a.

But how do we know that it is electron 1, and not electron 2, which is around nucleus A? Electrons carry no identification labels to show us. Indeed, the antisymmetry (Pauli) principle (p. 8) recognizes the fact that all electrons are identical. There is no conceivable way by which we could distinguish our original function $\phi_a(1)\phi_b(2)$ from another one $\phi_b(1)\phi_a(2)$, in which electron 1 is around B, and electron 2 around A. We may use the ideas discussed on p. 16 when describing the Ritz form of Rayleigh's principle, and argue that a proper wavefunction should have characteristics of both possibilities. We should then write†

$$\Phi = c_1\phi_a(1)\phi_b(2) + c_2\phi_b(1)\phi_a(2),$$

where the parameters c_1 and c_2 would have to be determined by the variation principle. We could of course carry through the argument in precisely this way. But it is unnecessary, since the equivalence of the two electrons implies that the two parts must play completely equal roles in the final wavefunction. Now in wave mechanics probabilities are determined by the square of a wavefunction, and if the interchange $1 \leftrightarrow 2$ is to make no difference this means that $c_1^2 = c_2^2$. So $c_1 = \pm c_2$, and if we do not bother at this stage with normalization the two allowed combinations can be written

$$\Phi_{\pm} - \phi_a(1)\phi_b(2) \pm \phi_b(1)_a(2). \tag{7}$$

These two functions are appropriate to the separated-atom situation in which the internuclear distance R is large. But Heitler and London suggested that they could also be used as approximate wavefunctions right into the molecular region of Fig. 8.

---

† We often use a capital letter, such as Φ, to denote a wavefunction that refers to more than one electron, keeping lower case letters for orbitals (i.e. *one*-electron wavefunctions).

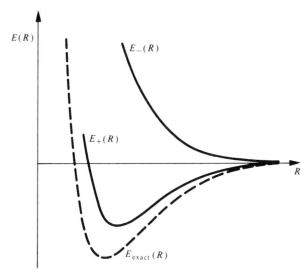

FIG. 9. The Heitler–London energies $E_{\pm}(R)$ for H_2, and the exact ground-state energy $E_{exact}(R)$.

Let us therefore substitute $\Phi_{\pm}$ in the Rayleigh ratio (5); we shall get two energies $E_{\pm}$ for each value of R. The calculations involve some very difficult integrals, whose mathematical evaluation had to wait for a further year after the original paper of Heitler and London. In the end two curves are obtained as shown in Fig. 9.

This is very satisfactory. It shows us that Φ_{+} leads to a stable molecule, whereas Φ_{-} does not. The one is an attractive state, the other is repulsive. From the valence point of view we are thus interested only in Φ_{+} and E_{+}. Unless there could be any doubt we shall henceforth drop the subscript $+$, and refer simply to Φ and E. The curve $E(R)$ is our approximation to the true PE curve for the molecule. It indicates an equilibrium bond length 87 pm to be compared with the experimental value 74 pm, and a dissociation energy D_e equal to 303 kJ mol^{-1} (3.14 eV) to be compared with the experimental value 458 kJ mol^{-1} (4.75 eV). Considering that we are using such a simple type of wavefunction with effectively no flexibility, the result is good.

We know that when using the Rayleigh ratio we shall always get an energy lying above the true value. Thus the true PE curve in Fig. 9 must lie everywhere below the approximate one, though as $R \to \infty$ with dissociation into two separate atoms, the calculated energy becomes nearer and nearer to the true one. So the PE curve must have a minimum; and the dissociation energy must exceed 303 kJ mol^{-1}.

We call Φ a valence bond (VB), or electron-pair, wavefunction. In pictorial terms we take an electron in an atomic orbital round A and 'pair' it with an electron in an atomic orbital round B. This combination of two electrons gives rise to the chemical bond. It could therefore be argued that this wavefunction transcribes into modern terms the ideas proposed in 1916 by G. N. Lewis, when he spoke of two electrons as being shared between the nuclei. This sharing is evidenced by the fact that in Φ we find electron 1 sometimes around nucleus A, in the form ϕ_a (1), and sometimes around B, in the form $\phi_b(1)$. It is this combination of the two parts in eqn (7) that leads to the PE curve of Fig. 9. For if we had used just the one term $\phi_a(1)\phi_b(2)$ we should have found a curve with hardly any indication of bonding at all. Heitler and London's fundamental contribution was to show that both parts are needed. Sometimes it is argued that in the two parts of Φ the electrons have changed places, and it is implied that electron exchange is a genuine phenomenon. It is nothing of the kind. The wavefunction (7) was derived for the separated atoms; when it is used for the molecular situation it becomes just one from among many possible trial functions. It is wrong, therefore, to attach physical significance to its precise form. In particular, there can be no operational significance in the claim that two electrons have exchanged places, and the Pauli principle itself rests on recognition of the fact that all electrons are perfectly indistinguishable.

A quantity which *does* have a certain physical significance may be obtained from any wavefunction Φ; it is the charge density discussed on p. 4. When this density, P, is computed from the wavefunction (p. 7), taking the upper sign, we obtain the contour map shown in Fig. 10. The peaks in electron density near the two nuclei are very apparent, and so is a building up of charge

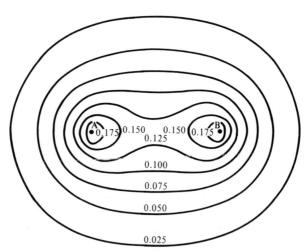

FIG. 10. Contours of total electronic charge-density for H_2. [The wavefunction used is a little more accurate than the simple Heitler–London function (7).] (From C. A. Coulson, *Proc. Camb. phil. Soc.* **34**, 204 (1938) by permission.)

between the nuclei. The general appearance of the charge-cloud, as seen 'from the outside', is thus similar to that of an ellipsoid of revolution. This represents the simplest picture of the shape and size of a chemical bond.

Possible refinements of the Heitler–London wavefunction

The Heitler–London wavefunction (7) is the simplest function of its kind, and there are literally hundreds of refinements, that different people have introduced in order to get a better wavefunction and energy. Two of them are worth mentioning.

In the first we recall the discussion of atomic helium in Chapter 1, and the way in which flexibility was introduced into a trial wavefunction by means of a variable orbital exponent in the AO. In the helium atom each electron felt a reduced attraction towards the nucleus, due to the 'screening' effect of the other electron. But in the hydrogen molecule there are *two* attracting centres and the net effect is an *increased* field pulling the electrons into the molecule. The resultant contraction of the charge cloud can be achieved if we use an atomic orbital $\sqrt{(c^3/\pi a_0^3)} \cdot e^{-cr/a_0}$ in which c differs from the hydrogenic value 1.0, and is to be regarded as a variable parameter. We do in fact find that at $R = R_e$, the value of c is about 1.2, indicating that the 'size' of the molecule is less than twice the 'size' of an atom, in agreement with conclusions from the kinetic theory of gases. The bond energy is also improved from 303 kJ mol^{-1} (3.14 eV) to 365 kJ mol^{-1} (3.78 eV), and R_e now assumes almost exactly the experimental value.

The second refinement is to argue that although at large nuclear separations it may be valid to say that the two electrons are always associated with different nuclei, it is no longer so at molecular distances. There is a possibility that both electrons may be around nucleus A. If they were, a suitable wave function would be $\phi_a(1)\phi_a(2)$. But then we argue, as before, that we must also include $\phi_b(1)\phi_b(2)$. Further, these two must play an equal part, by symmetry. Finally the correct symmetry combination to mix with (7) must be $\phi_a(1)\phi_a(2) + \phi_b(1)\phi_b(2)$. The Ritz method tells us to take a trial wavefunction

$$\Phi = \phi_a(1)\phi_b(2) + \phi_b(1)\phi_a(2) + \lambda\{\phi_a(1)\phi_a(2) + \phi_b(1)\phi_b(2)\} \qquad (8)$$

$$= \Phi_{cov} + \lambda\Phi_{ion}$$

where Φ_{cov} denotes the original covalent Heitler–London wavefunction with equal sharing of the two electrons, and Φ_{ion} denotes ionic terms in which both electrons are around the one or the other nucleus. This mixing of covalent and ionic terms is often described as 'resonance'. The parameter λ must be found variationally. At the equilibrium position $\lambda \approx \frac{1}{6}$, showing that for H_2 these terms are not very important. The PE minimum is lowered a bit further, so that now $D_e = 388$ kJ mol^{-1} (4.02 eV). Each new refinement leads us closer to the true value 458 kJ mol^{-1} (4.75 eV). We could go on making these refinements as long as we liked, but we should find it was increasingly difficult to lower the calculated energy sufficiently. However with 100 terms in a Ritz-type wave function Kolos and Wolniewicz† (1968) obtained results in complete agreement with experiment (p. 17).

The spin factor

At this point a word must be said about spin. So far we have considered only the spatial or 'orbital' part of the wavefunction; but from the discussion on p. 8 we should have assigned the electrons to *spin*-orbitals, $\phi_a(1)\alpha(1)$, for example, describing an 'up-spin' electron in orbital ϕ_a. The wavefunction would then be a mixture of eight possibilities, which are all equally acceptable provided we ignore the very small magnetic interactions between electron spins: the possible products are

$$\phi_a\alpha\phi_b\alpha, \quad \phi_a\beta\phi_b\alpha, \quad \phi_a\alpha\phi_b\beta, \quad \phi_a\beta\phi_b\beta$$

and in each product electron 1 may be assigned to the first spin-orbital, electron 2 to the second, or vice versa. Our equivalence argument again suggests that the wavefunction will be a mixture in which every product occurs with coefficient ± 1. But now we can also use the Pauli principle (p. 8); when we interchange the variables symbolized by 1 and 2, which describe the positions *and* spins of the two electrons, the wavefunction must change sign. It is easily verified that the correct combination of spin-orbital products corresponding to the orbital factor (7) with the plus sign is

$$\Psi_+(1, 2) = [\phi_a(1)\phi_b(2) + \phi_b(1)\phi_a(2)] \, [\alpha(1)\beta(2) - \beta(1)\alpha(2)]$$

Expansion shows this function to be a mixture of four of the eight spin-orbital products and the choice of signs ensures overall antisymmetry under the interchange $1 \leftrightarrow 2$. Admission of spin thus simply leads us from Φ_+ in (7) to the function

$$\Psi_+ = \Phi_+ \times \text{(antisymmetric spin function)} \tag{9a}$$

On the other hand, an equally acceptable wavefunction would be

$$\Psi_- = \Phi_- \times \text{(symmetric spin function)} \tag{9b}$$

in which there are *three* possible choices of symmetric spin factor $\alpha(1)\alpha(2)$, $\beta(1)\beta(2)$ or $\alpha(1)\beta(2) + \beta(1)\alpha(2)$. Since, to our level of approximation the spin does not effect the energy, Ψ_- must describe a *triplet* state of the molecule, *three* possible wavefunctions being associated with the one degenerate energy level, whereas Ψ_+ describes a *singlet* (i.e. non-degenerate) state. It may be shown that the antisymmetric spin function corresponds to zero resultant spin $S = 0$, the two spins $\frac{1}{2}$ being 'anti-parallel coupled' or 'paired', while the symmetric spin-function corresponds to $S = (\frac{1}{2} + \frac{1}{2}) = 1$ with the two spins 'parallel coupled'. It is important to realize, however, that the spin factor in (9a) or (9b) simply acts as an *indicator* which tells us (through our knowledge

† Kolos, W. *and* Wolniewicz, L. (1968). *J. Chem. Phys.* **49**, 404.

of the Pauli principle) whether the related orbital factor is symmetric or antisymmetric; the orbital factor alone serves to determine the energy and in the rest of this book we shall not be troubled by the spin.

Homonuclear diatomics

The general picture obtained for H_2 generalizes very easily to other homonuclear diatomic molecules. The bond, if it is a conventional single bond, will be due to the pairing of two electrons, initially in AOs round the two nuclei. The electrons will have opposed spins.

Consider diatomic lithium Li_2. We saw on p. 9 that the electronic structure of an isolated lithium atom is $(1s)^2(2s)$. The $(1s)^2$ pair of electrons are already spin-coupled, and cannot therefore be used to pair with anything else. In any case they are inner-shell electrons, and no use for bonding. So the bond makes use of the 2s AOs. The pairing of the two 2s electrons follows precisely the same pattern as with H_2. There is no need to set it out in detail. Since we use s-orbitals we may refer to the bond as an s-bond.

Next consider diatomic fluorine F_2. An isolated F atom has electronic configuration $(1s)^2(2s)^2(2p_x)^2(2p_y)^2(2p_z)$. All the electrons are spin-coupled except for the $2p_z$ electrons. So we must use these to form the bond. The wave function will formally be identical with that for H_2 and Li_2, except that now it will be a p-bond. However, before we can go any further we must look more closely at the geometrical relationships between the paired orbitals. We shall be led to a principle of the greatest importance.

Principle of maximum overlapping

We have already seen that according to the Heitler–London approximation the bonding in H_2 arises from the combination $\phi_a(1)\phi_b(2) + \phi_b(1)\phi_a(2)$. It does not arise from either of the two terms taken alone. This implies that the two terms 'interact' strongly with each other. It is natural to conclude that the effectiveness of this interaction depends on the two atomic orbitals ϕ_a and ϕ_b being reasonably close together. In more precise terms we require that the overlap of the two AOs, given by $S = \int \phi_a\phi_b \, d\tau$, shall be large. This is possible only if there are volume elements $d\tau$ in which the product $\phi_a\phi_b$ is not small: this can happen only if in these volume elements both ϕ_a and ϕ_b are reasonably large. We speak of an overlap region (Fig. 11). This region is not sharply defined, for the simple reason that each separate AO has no rigid boundary, as might be implied by the diagram, but stretches out to infinity in an exponentially decreasing fashion. We shall see later in this chapter that when two AOs overlap there will be a build-up of charge in the overlap region. The magnitude of this, as measured by the overlap integral S, is a good measure of the bonding that can be provided by the two AOs.

The importance of the overlap integral is confirmed by detailed analysis of contributions to the energy. Mixing of the two products in the

The overlap region

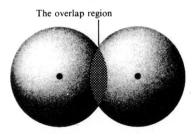

FIG. 11. The overlap region for two s atomic orbitals.

Heitler–London function (7) lowers the energy well below that expected for either product separately; and the extent of this bonding effect is dominated by a term arising from overlap.

This simple result has a profound influence on nearly all steric properties. If we bring together two hydrogen 1s AOs then, since each AO is spherically symmetrical, it does not matter (Fig. 11) in what direction they approach each other. But if we bring a 2p AO on atom A towards a 1s AO on atom B (Fig. 12) it makes a very big difference. If we bring them as in Fig. 12(a) there is maximum overlapping; but if we bring them as in Fig. 12(b) the overlap integral $S \equiv 0$, and no bonding results, on account of the difference in sign of a 2p orbital in the two lobes. Thus we can say that in this type of bonding a p-orbital is strongly directional.

In the case of F_2 dealt with in the last section it is now clear that the two p-orbitals must be directed towards each other, as in Fig. 12(c), if we are to get a good bond.

However, it is possible for the two p-orbitals to approach each other as in Fig. 12(d) where the directions of the two orbitals are parallel and at right angles to the internuclear axis. There is still an overlap region, so that bonding will result. But there is a considerable difference between the situations in (c) and (d). In the collinear case (c) the wavefunction has complete axial symmetry, as of course it also has in Fig. 11 and Fig. 12(a); in case (d) however, there is no such symmetry. For the plane shown through the nuclei is a nodal plane on which the wavefunction vanishes. If we were to rotate one AO around the nuclear axis we should soon reduce the overlap integral, so that the total energy would rise. Work would be needed to achieve this. In other words there is no restriction to rotation around the bond in cases (a) and (c), but there is in (d).

There is another overlapping pair of p-type AOs, entirely similar to (d) but with the directions of both p-orbitals turned through 90° around the axis.

When the bonding arises from a two-electron charge cloud with axial symmetry, we refer to it as being a single bond, and express the axial symmetry by the label σ-bond. This is the case for (a) and (c). When we have a situation such as (d) we call it a π-bond. Then a double bond will be a

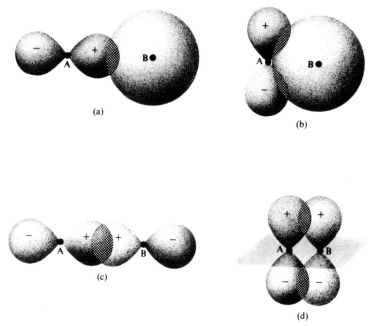

FIG. 12. Overlap between different atomic orbitals. (a) 2p of A with 1s of B (σ-type overlap). (b) 2p of A with 1s of B (zero by symmetry). (c) 2p of A with 2p of B (σ-type overlap). (d) 2p of A with 2p of B (π-type overlap).

combination of one σ-bond and one π-bond. If the direction of the two AOs in (d) is the x-axis, we could call it a π_x-bond. A triple bond would then be the combination of one σ-bond, one π_x-bond and one π_y-bond.

This is best illustrated by reference to molecular nitrogen, usually written N≡N. In Chapter 1 we saw that the ground state of an isolated nitrogen atom was $(1s)^2(2s)^2(2p_x)(2p_y)(2p_z)$. The orbitals that could be used for bonding are clearly the $2p_x$, $2p_y$ and $2p_z$ ones. Fig. 13 shows the approach of two such atoms. If the orbitals are located as shown we form a σ-bond by pairing together the orbitals labelled z_a, z_b; we form a π_x-bond by pairing x_a, x_b; and a π_y-bond by pairing y_a, y_b. The result is therefore the triple bond that we should have expected. A closer look at the wavefunction now shows that the overall charge distribution has axial symmetry, so that our description is not actually dependent on which directions we choose for x and y, provided only that they are at right angles to each other and to the internuclear axis.

One further point must be made. When we use the triple-bond symbol N≡N we are not now at liberty to say that all three bonds are equivalent. The π_x- and π_y-bonds are equivalent, but they differ from the σ-bond. Experimentally this is revealed by the fact that there are two different

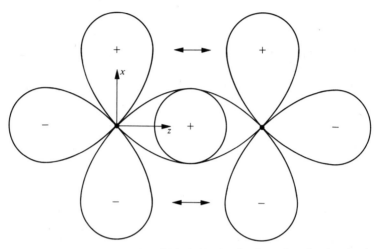

FIG. 13. Schematic representation of AOs in N_2, showing formation of a σ-bond and a π_x bond. The two $2p_y$ AOs, normal to the plane of the page, form a π_y bond. (See also Fig. 12 c, d.)

ionization energies, one arising from the removal of an electron from the σ-bond and another from the π-bond. We shall be dealing a bit later (p. 35) with the question of ionization energies, and shall therefore say no more now.

Heternuclear diatomics

We are now in a position to describe a bond between two dissimilar atoms. Let us take the case of HF as an example. When we recall that the isolated fluorine atom has a configuration $(1s)^2(2s)^2(2p_x)^2(2p_y)^2(2p_z)$ we see that, just as in F_2, the bond in HF must make use of the $2p_z$ atomic orbital of F. The hydrogen atom brings its 1s orbital. The situation is similar to that shown in Fig. 12(a), with the H nucleus lying along the axis of the $2p_z$ orbital of F. But now it is quite unrealistic to suppose that the covalent Heitler–London wavefunction will be adequate by itself. For we have plenty of chemical evidence that the fluorine atom is more electronegative than the hydrogen. This will lead to a resultant flow of charge from the hydrogen atom onto the fluorine atom. It means that in addition to the covalent Heitler–London function Φ_{cov}, which corresponds to equal sharing of the two electrons, the true wavefunction must also have characteristics symbolized by the charge distribution $H^+ F^-$. If we write Φ_{ion} for the corresponding wavefunction, and recall that the Ritz form of the variation method encourages us to mix together any set of functions that may describe one or other of the expected characteristics, we shall use a trial function

$$\Phi = \Phi_{cov} + \lambda\Phi_{ion}, \tag{10}$$

where λ is a parameter to be chosen such that the Rayleigh ratio (p. 14) is stationary. This form of Φ is almost identical with that used in eqn (8) for covalent–ionic resonance in H_2. It differs because now there is only one component in the ionic term, corresponding to both bonding electrons being around the fluorine atom, whereas in H_2 we had to consider both H^-H^+ and H^+H^-, giving them equal weights. Of course, if we wanted to do so, we could add to (10) another ionic term, to give $\Phi = \Phi_{cov} + \lambda\Phi_{ion} + \lambda'\Phi'_{ion}$, where λ' is another variable parameter, and Φ'_{ion} is the wavefunction representing H^-F^+. If we were to do this we should soon find that λ' was small, so small that except for very accurate wavefunctions it may be safely neglected. This is another example of the point made in Chapter 1, that the Rayleigh–Ritz method is ideally constructed to allow us to save time and effort by incorporating such chemical experience as we may have. Let us therefore stick to (10).

It would be very nice if we were able to calculate the value of λ without too much trouble. Unfortunately this is not possible, since at the very least we should have to include all the other electrons, not actually involved in the bond, in order to be sure that we take proper account of all the Coulomb electrostatic and exchange forces. For this reason attempts to calculate λ directly are usually made in some semi-empirical way, which we shall not describe, since there is a better way of estimating λ without any of these difficulties.[†] Following Pauling[‡] we appeal to the molecular dipole moment μ. It is a reasonable assumption (though not quite an accurate one!) to assume that the dipole moment arises solely from the two bonding electrons, and that (again not quite correct) the covalent function Φ_{cov} contributes nothing. Then the dipole moment is associated solely with Φ_{ion}.

TABLE 1

Ionic character in the hydrogen halides

Molecule	HF	HCl	HBr	HI
μ (debyes)[†]	1.82	1.03	0.83	0.45
λ (see eqn 11)	0.84	0.45	0.37	0.25

† 1 debye $= 3.334 \times 10^{30}$ C m.

We interpret the covalent–ionic function (10) as implying that the weights of the two parts are in the ratio $1^2 : \lambda^2$. Thus the weight of the ionic term is $\lambda^2/(1 + \lambda^2)$. This is called the *fractional ionic character* (FIC). Now the ionic

---

† These methods are described *CV*, Chapter 6.

‡ Pauling, L. (1960). *The nature of the chemical bond*, 3rd edition, Chapter 3. Cornell University Press, Ithaca, New York.

term by itself corresponds to unit charge being moved a distance R, so that it has a dipole moment eR. Taking into account the weight factor $\lambda^2/(1 + \lambda^2)$ we see that according to this model

$$\mu = eR \times \frac{\lambda^2}{1 + \lambda^2}. \tag{11}$$

If we had been able to calculate λ directly this would have enabled us to predict the dipole moment μ. Instead we work backwards. Using the observed μ and R we use eqn (11) to predict the value of λ. Table 1 shows the results obtained not only for HF but also for the other halogen hydrides. Similar accounts could be given for covalent–ionic resonance in other heteronuclear diatomic molecules.

The numerical values in Table 1 are very reasonable. As we should expect from chemical knowledge the electronegativity difference between the hydrogen and the halogen atom decreases as we go to heavier elements. So the λ values also decrease. The charge-clouds of these diatomic molecules also change systematically (a similar sort of change is illustrated in Fig. 22); but we shall subsequently return to this matter. We must first develop the alternative model for chemical bonding, i.e. the molecular-orbital method. We therefore turn to this now.

Molecular-orbital method: H₂

The Heitler–London valence-bond method just described depended on the assumption that a wave function valid in the separated-atom region of Fig. 8 could be extended to apply to the molecular region. But we could equally well have started with the united-atom region and extended a wave function designed for this region into the molecular region. This leads us to the molecular-orbital method, which is the strict counterpart for molecules of the atomic-orbital method for atoms.

It is instructive first to see this in terms of the relation between the hydrogen molecule H_2 and the united-atom helium (Fig. 14). If we start with He we suppose both electrons to have the same orbital function 1s, but opposed spins. Now imagine that we divide the nucleus with its charge $+2e$ into two equal parts $+e$, and slightly separate them. We should expect that we could still use the same type of description for the electrons, though the orbitals would now be a little elongated in the direction of the molecular axis. If we now separate the two positive charges to chemical-bond distances we may still hope to use this model. The bond is then described by two electrons with paired spins, but in identical orbitals. From the nature of these molecular orbitals they will have axial symmetry around the nuclear axis. We call them σ-type. So the molecular description is $(1\sigma)^2$. Molecular orbitals for diatomic molecules are therefore bicentric, as compared with monocentric atomic orbitals.

We must look a little more closely at these MOs. We are considering the

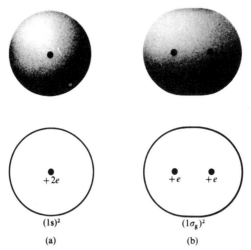

FIG. 14. Relation between atomic orbitals in He (a) and molecular orbitals in H_2 (b). Upper diagrams represent charge density in each atomic or molecular orbital; lower diagrams represent the above relation schematically.

wavefunction for one electron at a time, in the presence of the two nuclei and the other electron. When this one electron is near nucleus A the dominant forces on it will be similar to those that it would experience if the atom were isolated. So locally the wave equation will bear some resemblance to the corresponding equation for an electron in atom A. Thus the wavefunction near A will resemble an atomic orbital ϕ_a of atom A. Similarly near B the MO will resemble an atomic orbital (ϕ_b) of atom B. The Ritz principle now leads to the view that the complete MO, which we shall call ψ, might be written

$$\psi = c_1\phi_a + c_2\phi_b,$$

where c_1 and c_2 are variable parameters. We could find their values by the use of the variation method, and in a general case we should have to do this. But in the case of H_2 we can once again use symmetry. This time it is symmetry with respect to interchange of A and B: it forces us to put $c_1^2 = c_2^2$, so ensuring the symmetry of the charge density ψ^2. Thus there are two MOs whose unnormalized forms are

$$\psi_\pm = \phi_a \pm \phi_b. \tag{12}$$

For the ground state of H_2 we take ϕ_a and ϕ_b to be the 1s AOs. The reader will not be surprised to learn that the $+$ sign corresponds to a lowering of the energy, and the $-$ sign to an increase. We refer to these as the bonding and antibonding combinations of ϕ_a and ϕ_b. In the ground state it is natural to put both electrons in ψ_+. When we evaluate the total molecular energy we find a curve similar to the Heitler–London curve in Fig. 9, except that the calculated D_e is a little worse, being 260 kJ mol^{-1} (2.70 eV) instead of 303 kJ mol^{-1} (3.14 eV).

We can of course improve this wavefunction just as we improved the primitive Heitler–London wavefunction. If, as on p. 17, we allow for a variable orbital exponent c, we again find a value similar to that with the Heitler–London function, and a bond energy 337 kJ mol^{-1} (3.49 eV). There are many other improvements that can be incorporated if we have the time and energy to do so. Our conclusion is that the MO method can be used with about the same success as the VB method to describe the bond in molecular hydrogen. In fact it can be shown that when we extend the two methods appropriately they become identical. There is thus no reason for preferring the one to the other (except perhaps for excited states where the MO method is usually much simpler). This equivalence is easily shown in detail for H_2, and can be proved for more complicated many-electron systems. In our own qualitative study of chemical bonding in the rest of this book, we shall use both methods. The VB method sometimes seems more 'chemical', but if we were wanting to make very accurate numerical calculations experience suggests very strongly that the MO method, with its various extensions, would be the better one to use.

Homonuclear diatomic molecules

According to eqn (12) the lowest-energy MOs for H_2 are of the form $\phi_a \pm \phi_b$ where ϕ_a and ϕ_b are 1s AOs on the two nuclei. However it is clear that the arguments which we used in coming to this wavefunction, being wholly based on symmetry considerations, were independent of the choice of AO, provided only that we used similar orbitals on both nuclei. This means that we can start with any AO on nucleus A and combine it with a similar AO on nucleus B, to give a bonding and an antibonding MO. We sometimes represent this as in Fig. 15.

In this way by starting with each atomic orbital on one atom we get approximations to two MOs for the homonuclear diatomic. But before we can discuss them properly and use the *Aufbau* principle we need to say more about their symmetry properties. As always in quantum theory each type of symmetry in the molecule, reflected as it must be in the symmetry of the Hamiltonian, will lead to a corresponding symmetry in the orbital.

In a homonuclear diatomic molecule there are three fundamental symmetry operations which leave the Hamiltonian unchanged. Any other such symmetry operations will be merely some combination of these three. The fundamental symmetries are:

   (i) rotation symmetry around the molecular axis,
   (ii) inversion in the centre of the molecule, and
   (iii) reflection in a plane containing the molecular axis.

Condition (i) implies that MOs behave in a particular way under rotation around the axis; they may have complete axial symmetry (σ-type), or may

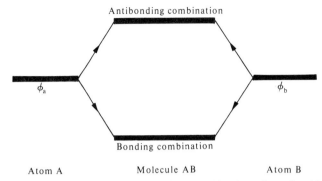

FIG. 15. Formation of bonding and antibonding combinations of atomic orbitals ϕ_a and ϕ_b to form molecular orbitals.

have one angular node (π-type), as in Fig. 12(d), or perhaps two nodes (δ-type), and so on. The symbols $\sigma, \pi, \delta, \ldots$ are the molecular analogues of s, p, d, $\ldots$ for AOs. Condition (ii) implies that each MO is either even or odd with respect to this inversion. Just as in atoms we use the labels g (German *gerade*, or even) and u (*ungerade*, or odd). Condition (iii) implies that the MO is either even or odd with respect to this reflection; we use the superscript $\pm$. Frequently the last symbol is omitted as it is usually obvious which symmetry is involved. Finally we have the quantum number that orders each class of MOs according to their sequence of energies.†

It follows from this that the two MOs (12) involving 1s AOs will be labelled $1\sigma_g$ and $1\sigma_u$. The ground state electron configuration of H_2 will be $(1\sigma_g)^2$ and the state itself is described as $^1\Sigma_g^+$, where capital Greek letters $\Sigma, \Pi, \Delta, \ldots$ for the complete wavefunction are analogous to $\sigma, \pi, \delta, \ldots$ for the individual MOs.

In order to deal with homonuclear diatomics we need to know the sequence of energies, in much the same way that it was needed in Fig. 5 to deal with atoms. The sequence most usually found is

$$1\sigma_g < 1\sigma_u < 2\sigma_g < 2\sigma_u < 1\pi_{xu} = 1\pi_{yu} < 3\sigma_g < 1\pi_{xg} = 1\pi_{yg} < 3\sigma_u < \ldots \quad (13)$$

but sometimes the $3\sigma_g$ lies below the $1\pi_u$ degenerate pair. The way in which these MO energies are related to those of the AOs out of which they are composed is shown in Fig. 16.

† At different times many different notations have been used. Thus the quantum number has sometimes been related to the principal quantum number of the united-atom form of the MO, and sometimes to the separated-atom component ϕ_a or ϕ_b. Sometimes an asterisk * has been used to denote an antibonding combination. The notation used here and later has now become almost standard.

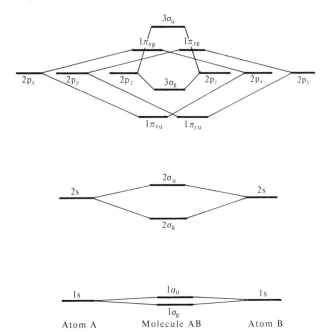

FIG. 16. Formation of molecular orbitals for homonuclear diatomic molecules from the original atomic orbitals of the two atoms (slightly simplified version).

It is now a simple matter to set down the electronic configuration of the ground state of molecules of this class. A few examples are shown below

H_2 $(1\sigma_g)^2$ $^1\Sigma_g$

He_2 $(1\sigma_g)^2(1\sigma_u)^2$ $^1\Sigma_g$

Li_2 $(1\sigma_g)^2(1\sigma_u)^2(2\sigma_g)^2$ $^1\Sigma_g$

N_2 $(1\sigma_g)^2(1\sigma_u)^2(2\sigma_g)^2(2\sigma_u)^2(1\pi_{xu})^2(1\pi_{yu})^2(3\sigma_g)^2$ $^1\Sigma_g^+$

O_2 [$\quad\ldots\ldots\quad$] $(1\pi_{xg})(1\pi_{yg})$ $^3\Sigma_g^-$

F_2 [$\quad\ldots\ldots\quad$] $(1\pi_{xg})^2(1\pi_{yg})^2$ $^1\Sigma_g^+$

Ne_2 [$\quad\ldots\ldots\quad$] $\ldots\ldots$] $(3\sigma_u)^2$ $^1\Sigma_g^+$

In the case of O_2 where two electrons have to be fitted into the degenerate pair of MOs $1\pi_{xg}$, $1\pi_{yg}$ we make use of Hund's rules (p. 9) to the effect that the electrons will go into different orbitals, and moreover will have parallel spin. This leads to a paramagnetic ground state, in excellent agreement with experiment. Historically it was the natural way in which this paramagnetic state was to be expected that played a large part in getting the MO method accepted.

The filling of a bonding MO and its antibonding partner leads to zero bonding: more accurately it leads to slight antibonding. Thus we see that He_2 and Ne_2 should be unstable in their ground states, though some of their excited states and also their positive ions are stable entities. In H_2, Li_2, F_2 it appears that the bonding is due to just two electrons in some appropriate MO. It is natural to say that we are dealing with a single bond. Further the electrons in question occupy σ-type orbitals. These are therefore σ-bonds. In N_2 there are five MOs, each doubly occupied by valence-shell electrons. The bonding and antibonding effects of the $2\sigma_g$ and $2\sigma_u$ MOs effectively cancel, leaving us with a net result of three pairs of bonding orbitals. So we have a triple bond, and, just as in the VB scheme, this is seen to consist of one σ-bond and two π-bonds.

Ionization potentials

Thanks to the work of D. W. Turner and W. C. Price it is now possible to confirm by experiment the accuracy of the MO configurational description just given. If a homogeneous beam of light of known frequency v is allowed to fall on one of these molecules, some of its energy may be used in photoelectric emission, or ionization, of an electron. If the energy of the ejected electron is E_e, then the ionization energy I will be given from the conservation of energy:

$$hv = E_e + I.$$

By measuring the energy of electron emission a series of peaks is found at well defined E_e values: and hence I is determined.

The case of N_2 is shown in Fig. 17. According to our description of N_2 there should be four distinct valence-electron ionization potentials, corresponding to removal of an electron from a $2\sigma_g$, $2\sigma_u$, $1\pi_u$, or $3\sigma_g$ MO. All four of these are shown experimentally. From a careful study of the four peaks it is possible to identify the symmetry of the relevant MOs, and hence to confirm the sequence of energies which we have given. It is a rather remarkable situation that R. S. Mulliken and others had used MO theory to predict the type and number of molecular orbitals for about thirty years before they were fully confirmed by experiments such as these. MO theory is now well-established as a powerful tool for understanding ionization processes.

Heteronuclear diatomic molecules

At this stage it is not difficult to extend our discussion to deal with heteronuclear diatomic molecules. We shall continue to use the *Aufbau* principle. The chief new features are that: (i) if the nuclei of atoms A and B are different we shall no longer have the convenient u, g symmetry associated with inversion; but of course we shall still retain the σ, π, δ, ... classification

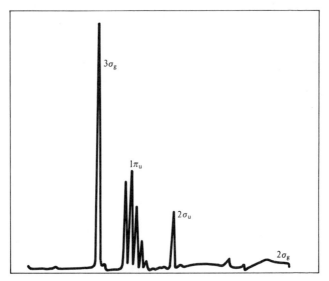

FIG. 17. Photoelectron spectrum of N_2 (courtesy of W. C. Price), showing ionization from the four distinct occupied valence-shell molecular orbitals. The structure of the peaks arises from vibrational effects.

and also the $\pm$ symmetry associated with reflection in a plane containing the nuclei; and (ii) each MO will now be built up from different AOs on the two nuclei, rather than from identical ones.[†] The criteria to be used will still be that there should be maximum overlapping and also that the energies of the two AOs in their separate atoms must be roughly equal. This latter criterion implies that the component AOs shall have roughly the same size. However, the fact that the component AOs are different will mean that we must now introduce a coefficient of mixing, writing

$$\psi = c_1 \phi_a + c_2 \phi_b \tag{14}$$

where c_1^2 is no longer equal to c_2^2. The relative magnitudes of c_1 and c_2 should in principle be found from the variation principle, just as in other situations previously dealt with. However, it should also be clear that since one interpretation of eqn (14) is that for an electron in this MO a fraction $c_1^2/(c_1^2 + c_2^2)$ is associated with atom A and $c_2^2/(c_1^2 + c_2^2)$ with atom B, we could work backwards from the observed dipole moment μ, just as with the VB model earlier in this chapter.[‡] If we can neglect any dipole contributions from

[†] This is the LCAO approximation (Linear Combination of Atomic Orbitals) due to Mulliken, of which we shall make much more use in Chapter 5.

[‡] In some cases the distribution of the electrons in a molecule can be determined by electron spin resonance or nuclear magnetic resonance: this is described by K. A. McLauchlan in *Magnetic resonance* (OCS 1).

any other electrons, then the bond represented by ψ^2 should have a dipole moment $\mu = (c_1^2 - c_2^2)(c_1^2 + c_2^2) \times 2eR$. We shall not list any detailed numbers at this stage because they follow precisely the same pattern as was previously discussed (Table 1) for the VB scheme.

The case of HF is worth presenting in a little more detail, since it illustrates the preceding discussion, and may be compared with the VB account on p. 25. We start off as in Fig. 18 by listing, on the right and left sides of the diagram, the AOs of H and of F, together with their energies. It is immediately apparent that the 1s orbital of hydrogen has an energy which is close only to that of the 2p AO of fluorine. The other AOs of fluorine will this remain effectively unchanged on passing to the molecule. Now the 2p orbital is triply degenerate. Moreover the $2p_x$ and $2p_y$ members have π-symmetry (if, as usual, we take the molecular axis to be in the z-direction). So, since the hydrogen 1s orbital has σ-symmetry, it cannot mix with either of these, which will therefore remain unchanged, and can be called lone-pair, or non-bonding,

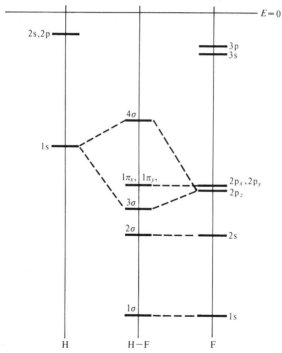

FIG. 18. Formation of molecular orbitals in HF from the atomic orbitals of H and of F. In the ground state all the MOs shown are doubly occupied except 4σ, which is empty.

orbitals. The bond arises from two electrons in a MO ψ of the form

$$\psi = c_1 H(1s) + c_2 F(2p_z). \tag{15}$$

There is both a bonding and an antibonding mixture. Both are shown in the diagram.

Using proper MO language this means that the ground state of HF is

$$(1\sigma)^2 (2\sigma)^2 (3\sigma)^2 (1\pi_x)^2 (1\pi_y)^2.$$

Since the bond arises from the 3σ orbital, it is a normal single bond. Further, since fluorine is more electronegative than hydrogen we expect c_2 to be bigger than c_1 in the representation (15). This leads to a dipole moment in the correct direction $H^+ - F^-$. If we use the experimental value $\mu = 1.82$ debye for the moment, then c_2/c_1 in eqn (15) is about 1.80. Since probabilities are proportional to squares of wavefunctions and $1.80^2 = 3.24$, this shows just how much the bonding orbital is concentrated at the fluorine end of the bond rather than at the hydrogen end.

Charge distribution in molecular hydrogen

It is interesting to compare the VB and MO wavefunctions in terms of their predicted charge distributions. The MO expression is easily found. In general (p. 4) each occupied orbital ψ makes its own contribution ψ^2 (or $2\psi^2$ if *doubly* occupied) to the total charge density. The orbitals we use for this purpose must be normalized so that $\int \psi^2 \, d\tau = 1$; the combination $\phi_a + \phi_b$ does not have this property until it is multiplied by a normalizing factor. We can easily verify that

$$\psi = \frac{\phi_a + \phi_b}{\sqrt{(2(1 + S))}}$$

(where S is the overlap integral $\int \phi_a \phi_b \, d\tau$) is properly normalized and it follows at once that the electron density, with 2 electrons in this bonding MO is

$$P = 2\psi^2 = q_a \phi_a^2 + q_b \phi_b^2 + 2p_{ab} \phi_a \phi_b \tag{16}$$

where

$$q_a = q_b = p_{ab} = \frac{1}{1 + S} \tag{17}$$

The first two terms in (16) are simply AO densities multiplied by numerical 'weight factors' (q_a, q_b) which indicate their relative importance; as the overlap (Fig. 11) is positive each hydrogen-like density occurs with a weight somewhat smaller than in the free atom. The third term in (16) is appreciable only where *both* ϕ_a and ϕ_b are large i.e. *in the region of overlap*; the numerical factor p_{ab} therefore tells us something about the degree of *bonding*. Of course, if we integrate the density (electrons/unit volume) over all space we must get the number of electrons—in this case 2, of which q_a come from the AO region ϕ_a^2, q_b come from ϕ_b^2, and $2p_{ab}S$ come from the overlap region defined by $\phi_a\phi_b$. Thus

$$q_a + q_b + 2p_{ab}S = 2$$

and the three terms indicate the amounts of charge (in electrons) associated with the three regions of space; q_a and q_b are known as 'charges' (or as 'populations' of ϕ_a and ϕ_b), while p_{ab} is a 'bond order'. We return to these ideas later.

For more elaborate wavefunctions, the charge density expression may be harder to obtain, but provided we build our wavefunction out of AOs it always takes the same form—only the numerical coeficients of the AOs change. Exactly the same expression (16) therefore applies to the VB functions used in earlier sections; we need only change the values of q_a, q_b, and p_{ab}. The wavefunction Φ_+ in (7) leads to†

$$q_a = q_b = \frac{1}{1 + S^2}, \qquad\qquad q_{ab} = \frac{2S^2}{1 + S^2} \qquad (18)$$

and these estimates are changed again if ionic terms are added as in (8).

The effect of bonding may be described by subtracting from P the charge density corresponding to the simple superposition of the densities of the free atoms to obtain a 'difference density' ΔP. For H_2, for example, we should subtract $\phi_a^2 + \phi_b^2$ for two separate hydrogen atoms located in their molecular positions. It is then easy to show that ΔP is positive around the centre of the molecule and negative on the farther sides of each nucleus. Thus the establishment of the bond is associated with a 'sucking' of electron density from the farther parts of the molecule into the overlap region between the nuclei, just as would be expected intuitively. If we plot the final density along the molecular axis we get the curve shown in Fig. 19. This illustrates the build-up of charge between the nuclei, and is typical of normal covalent bonds. The curve in Fig. 19 has almost the same appearance both for the MO

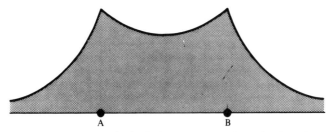

FIG. 19. Total electronic density in H_2 at points on the molecular axis AB.

† A full discussion is given in CV, Section 5.4.

and VB densities, with a slightly bigger build-up predicted for the MO than for the VB wavefunction.

Chemical crystallographers now commonly represent the effect of bond formation by plotting contours of the difference density. Experimentally this is obtained by subtracting from the measured molecular density the sum of atomic densities located at the positions of the nuclei. The difference-density contour diagram for H_2 is shown in Fig. 20. Since the total charge in a difference-density diagram must be zero, some regions (where there is a build-up of charge) will be positive, others (where there is a loss of charge) will be negative. The corresponding contours are usually drawn as continuous lines (increase of electron density) or dotted lines (decrease of density).

The picture shown in Figure 20 for molecular hydrogen is typical. Figure 21 shows the corresponding difference density for diatomic lithium.

It will be asked: how much charge moves into the overlap region in this way? The answer is: usually between 0.1 and 0.3 of an electron. This is not much, and its smallness explains why, until recently, the experimental evidence for it was not very good. The difficulty can soon be recognized if we look for this overlap charge in the C—C bond of ethane. Since there are eighteen electrons in C_2H_6 we are looking for about one per cent of the total charge. Only when high-resolution X-ray measurements have been made, and great care is taken with corrections for extinction and vibrational and rotational motion, is it possible to achieve this accuracy. But it *is* now possible. One very good example is that of diamond, where we can think of a single crystal as composed of a very large number of C—C bonds similar in many respects to the C—C bond of ethane. Dawson† has shown that about one tenth of an electron per bond is redistributed in this way. It is very satisfactory that one of the first theoretical predictions about the charge distribution in a covalent bond should now have a complete experimental confirmation.

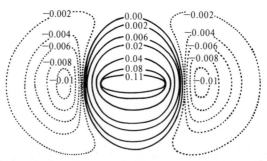

FIG. 20. Difference-density contours in H_2. Solid lines denote excess electronic charge, dotted lines denote decrease in density. (Courtesy of R. W. F. Bader.)

† Dawson, B. (1967). *Proc. R. Soc.* A **298**, 264.

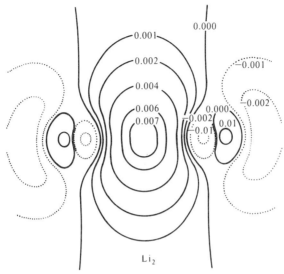

FIG. 21. Difference-density contours in Li_2. Notice the close general similarity with Fig. 20 for H_2. Both have the same kind of bond, in which charge is sucked from the two far sides of the molecule into the overlap region between the nuclei. (*Science* **151**, 961 (1966) courtesy of A. C. Wahl.)

The situation with ionic bonding is rather different. For now charge is actually transferred from one atom to the other. This follows from the VB wavefunction (10) on account of the ionic term in $\Phi = \Phi_{cov} + \lambda\Phi_{ion}$. In the MO wavefunction (14) it is revealed by the difference in the coefficients c_1, c_2 in the MO $\psi = c_1\phi_a + c_2\phi_b$. If, as in HF, the transfer of charge is large (large ionic character) then we may almost lose trace of that part of the bonding electron round the more electropositive atom. All this is very beautifully illustrated in Fig. 22, which shows the contours of charge density for the monohydrides of the series of atoms Li, Be, . . . F. It is interesting that in LiH, the greater electronegativity of H as compared with Li leads to a flow of charge onto the hydrogen atom and a dipole of direction $Li^+—H^-$. In BeH the bond is almost covalent, with very small dipole moment. But for the heavier hydrides charge flows away from the hydrogen, so that by the time we have got to HF there seems hardly any evidence of the original hydrogen-atom charge around the proton. Now, of course, the dipole has direction $F^-—H^+$.

We could say that the series of diagrams in Fig. 22 illustrates the change in direction and magnitude of the ionic part in the covalent–ionic resonance wave function. The changes in shape of the bond follow naturally.

The conclusion that we draw from this is that, whether we adopt a VB or a MO type of wavefunction, a single bond between two atoms involves a pair of

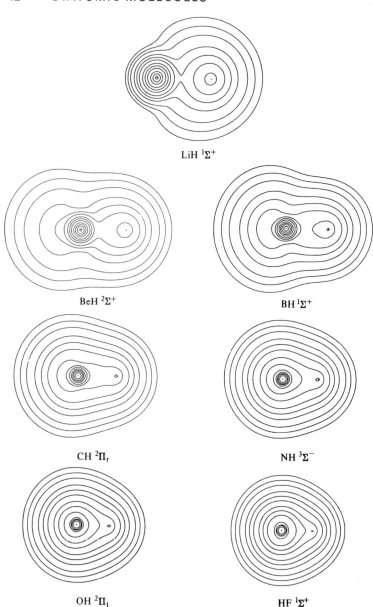

LiH $^1\Sigma^+$

BeH $^2\Sigma^+$ BH $^1\Sigma^+$

CH $^2\Pi_r$ NH $^3\Sigma^-$

OH $^2\Pi_1$ HF $^1\Sigma^+$

FIG. 22. The mono-hydrides of the atoms from Li to F. In each case the H atom is on the right. Notice how the charge contours indicate the changing role of the ionic part of the wave function. (Courtesy of R. W. F. Bader.)

electrons with opposed spins. The wavefunction has axial symmetry and is related to the atomic orbitals occupied by the two electrons in the separate atoms. Covalent bonding shifts electron density into the overlap region; ionic bonding shifts it from one atom to the other. We shall find that a similar description is possible for the bonds in polyatomic molecules.

PROBLEMS

2.1. Use eqns (17) and (18) to show that since $0 < S < 1$ the simple MO model for H_2 indicates a greater piling-up of charge in the overlap region than does the simple VB model.

2.2. Assuming that in diatomic He_2 the electronic description is $(1\sigma_g)^2(1\sigma_u)^2$, where the two MOs are given by eqn (12), show that there is a net loss of electrons from the overlap region, and hence that the molecule is unstable. {Hint: show that the two normalized MOs are $(\phi_a \pm \phi_b)/(2 \pm 2S)^{\frac{1}{2}}$.}

2.3. Write down the electronic configurations for the ground states of: (i) Na_2, (ii) K_2, (iii) LiH, (iv) LiF.

2.4. In which of the diatomic molecules N_2, O_2, and F_2 would you expect that ionization of the least-tightly-bound electron would lead to an increase in the bond energy? (Hint: does the ionization remove a bonding or an antibonding electron?).

2.5. How many distinct ionization energies should there be for the isolated HF molecule?

3. Polyatomic molecules

Bond properties

IT is an experimental fact that many bonds have very nearly constant properties. Thus the O—H bond has a length that hardly changes from 96 pm when we turn from water H—O—H to methanol CH_3—O—H. Tables of bond properties are available in many textbooks of physical chemistry. These properties include bond length, force constant, dipole moment, polarizability etc. It is chiefly because of this near-constancy of so many bond properties that the concept of a chemical bond is so valuable.

But at first sight these characteristic bond properties present a problem. In water, for example, there are two inner-shell electrons close to the oxygen nucleus, and a total of eight other electrons. We may call these the valence-shell electrons. Whatever bond properties there may be must be due to the distribution of these eight electrons. The existence of bond properties almost forces us to conclude that the total electronic charge can be approximately partitioned into pairs of electrons nearly localized in the region of each bond. In H_2O, therefore, we must suppose that two electrons 'constitute' each O—H bond, leaving four non-bonding electrons around the oxygen nucleus.

How should we attempt to describe these localized charge distributions? It is natural to carry over the discussion in Chapter 2 dealing with diatomic molecules, and treat each bond in a polyatomic molecule as if it could be described by two electrons. Normally one electron will be provided by each of the two atoms joined by the bond. The bond itself will have a 'personal wavefunction'; and this wavefunction will be built out of atomic orbitals of the two atoms. Moreover the two electrons will have opposed (i.e. paired) spins, and the relevant AOs must be chosen so that they overlap as much as possible.

Let us apply this idea to the water molecule. The isolated oxygen atom has an electronic structure $(1s)^2(2s)^2(2p_x)(2p_y)(2p_z)^2$. If we are to form electron pairs with the two hydrogen electrons we must not attempt to use any electrons that are already coupled together within the oxygen atom. For the Pauli principle would soon lead us into trouble if we did try to do so. We are left with just two available AOs on the oxygen atom: they are the $2p_x$ and $2p_y$ orbitals. This immediately tells us that oxygen is normally divalent, and can form two bonds. Indeed we have the general result that the normal valence number of an atom is equal to the number of unpaired valence-shell electrons in the atom. Now (Fig. 23(a)) the $2p_x$ and $2p_y$ orbitals will point along the x- and y-directions. To get maximum overlapping between these orbitals and the hydrogen 1s orbitals we must put the two hydrogen atoms along the x- and y-directions. We can then form two localized bonds as shown

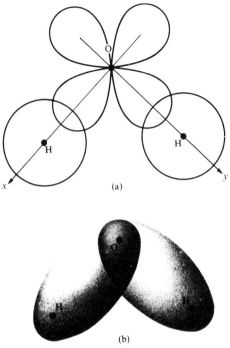

FIG. 23. Formation of localized O—H bonds in H_2O. (a) The isolated atomic orbitals before bond formation (but see p. 47). (b) Schematic representation of the two bonds.

schematically in Fig. 23(b). Each of these will have a personal wavefunction, either of MO or VB type, just as described in the previous chapter. If we adopt the Heitler–London model, we shall describe each O—H bond in terms of covalent and ionic contributions: and, basically, each of these will be built up from the appropriate O(2p) and H(1s) orbitals.

One immediate result of the greatest importance follows from this discussion. The valence angle should be approximately 90°, leading to a molecular shape that is triangular and not linear. The fundamental reason for this is that we had to use p_x and p_y orbitals of oxygen (or, alternatively, any pair of p-orbitals that were directed at right angles), since if the bonds are to be distinct (see p. 7) their wavefunctions must be orthogonal.

We need to discuss this orthogonality of p-orbitals a little more. Consider (Fig. 24) an arbitrary p-orbital, p, whose axis lies in the xy-plane and makes an angle θ with p_x. In general p and p_x are not orthogonal. It can be shown in fact that a p-orbital in any given direction can be resolved, just like a vector, into the sum of two component p-orbitals in perpendicular directions.†Thus

† *CV*, Chapter 7.

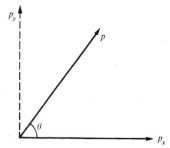

FIG. 24. The overlap integral between two p-orbitals, p and p_x.

p could be resolved into $\cos \theta \times p_x$ plus $\sin \theta \times p_y$ and the overlap integral $\int p_x p \, d\tau$ has the value

$$\int p_x (p_x \cos \theta + p_y \sin \theta) \, d\tau = \cos \theta \int p_x^2 \, d\tau + \sin \theta \int p_x p_y \, d\tau.$$

Now if the orbitals are normalized $\int p_x^2 \, d\tau = 1$, and $\int p_x p_y \, d\tau = 0$. So the required overlap integral is just $\cos \theta$. It vanishes if, and only if, $\theta = 90°$, i.e. if the atomic orbitals are at right angles. Of course, the choice of x- and y-directions is arbitrary. *Any* two p-orbitals will be orthogonal only if they are inclined at $90°$; otherwise they have an overlap $\int p_1 p_2 \, d\tau = \cos \theta_{12}$.

Hybridization

Up to this moment we have always supposed that the AOs to be used, whether in MO or VB models, are pure s- or p-orbitals. We must now ask whether this is really necessary. It is clear that s, p, d, ... orbitals arise naturally when, as with an isolated atom, there is one fixed centre of force around which the electrons move. But it is much less clear that we should restrict ourselves to them when, as in a molecule, there is more than one attractive centre.

Let us look more closely at diatomic lithium Li_2. The isolated lithium atom has electronic configuration $(1s)^2$ $(2s)$, and it was therefore very natural, in Chapter 2, that we should treat the Li—Li bond as if it were a pure s-bond. However, there is a 2p orbital not very much higher in energy (about $2\frac{1}{2}$ eV). So we might expect that, under the influence of the other atom to which it is bonded, the 2p orbital of the one atom should play some part in the bond. When forming the orbitals ϕ_a, ϕ_b on the two atoms, which will be used in the full molecular wavefunction, we should not be justified in treating them as pure s or p, but as mixtures, or hybrids, of s and p. Let us therefore look at the nature of a hybrid $\phi = s + \lambda p$, where λ is some numerical parameter that measures the mixing of s and p. We could interpret ϕ as implying that s and p occur in the ratio $1:\lambda^2$. If $\lambda = 0$ we have pure s; and if $\lambda \to \infty$ we have pure p. If λ lies between these extreme values we have a hybrid whose general form is represented schematically in Fig. 25. This is because the two lobes of a p-orbital are associated with opposite signs for the wavefunction. Thus in a combination $s + \lambda p$ the two components add together positively in the one direction and negatively in the other. The result is a hybrid with much stronger directional character than either a pure s or a pure p alone. It is therefore much better suited to good overlapping, and hence to an increase in the overlap energy. However we must be careful before concluding that λ is of the order of unity. This is because if we include a large amount of p-character in the hybrid, we are increasing the purely atomic energy on account of the greater energy of p than of s (see Fig. 5). In fact there is a compromise. We gain overlap energy by introducing p-character into the previous 2s AO of lithium, but we lose atomic energy. In the case of Li_2, calculations suggest that the bonding orbital is approximately 85 per cent of 2s and 15 per cent of 2p character.

There is no reason why this kind of argument should not be applied to other situations, including, of course, the diatomic molecules discussed in Chapter 2. Its importance, however, is greatest for polyatomic molecules. So let us return once more to water H_2O. Instead of making each O—H bond out of an O(2p) and an H(1s) orbital, we could get better overlapping if we used a combination $p + \lambda s$ for the oxygen AO. But the price to be paid for this better overlapping is that we have broken up the previous closed sub-shell $O(2s^2)$ by using part of the 2s orbital in each of the two O—H bonds. This costs energy. The effect is important if the resulting gain outweighs the cost. In the case of H_2O this is the case: and so the O—H bonds make use of hydrogen 1s AOs at the hydrogen end, and hybrids of s and p at the oxygen end.

This situation has a most interesting corollary in terms of the HOH valence angle. We have already seen that the two hybrids at the oxygen atom must be orthogonal. But these similar hybrids $p_1 + \lambda s, p_2 + \lambda s$ can only be orthogonal if the angle θ_{12} between them (Fig. 26) is such that

$$\cos \theta_{12} = -\lambda^2. \qquad (19)$$

This follows from the fact that

$$\int (p_1 + \lambda s)(p_2 + \lambda s)\, d\tau = \int p_1 p_2\, d\tau + \lambda \int s p_1\, d\tau + \lambda \int s p_2\, d\tau + \lambda^2 \int s^2\, d\tau.$$

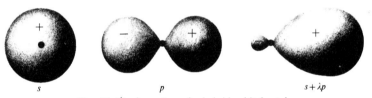

s   p $s + \lambda p$

FIG. 25. The formation of a hybrid orbital $s + \lambda p$.

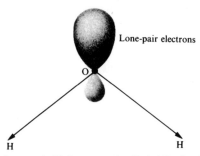

Lone-pair electrons

FIG. 26. Lone-pair electrons in H_2O as a result of hybridization in the O—H bonds.

The first integral on the right-hand side has already been shown (p. 46) to have the value $\cos \theta_{12}$. The second and third integrals are zero, and the final term is just λ^2. So the right-hand side is equal to $\cos \theta_{12} + \lambda^2$. Orthogonality demands that this vanish. Hence $\cos \theta_{12} = -\lambda^2$, as required. It follows that the directions of the two hybrids, and hence of the two O—H bonds, can no longer be at right angles, but, since $\cos \theta_{12}$ is negative, θ_{12} itself must exceed 90°. Hybridization therefore has the effect of opening out the H_2O valence angle.

There is yet one more important corollary for the water molecule. It springs out of the simple fact that by using part of the 2s AO of oxygen in forming the hybrids for the bonds, and by leaving part of the $2p_x$ and $2p_y$ AOs unused, we compel the two electrons previously in the orbital 2s to be in a hybrid orbital involving what is left of the original 2s, $2p_x$, and $2p_y$ AOs. All three hybrids have to be orthogonal. This can easily be shown to be possible. As a result the two electrons in the third hybrid have a charge density shown schematically in Fig. 26. We call them *lone-pair* electrons. But they have an off-centre distribution which, together with the other lone-pair electrons in the unchanged oxygen $2p_z$ AO, is tremendously important in biological behaviour. They provide the mechanism for hydrogen bonding, and base-pairing in the double strand of the DNA helix.†

Methane

The fundamental formula (19) relating the degree of s–p mixing in two equivalent hybrids to the angle between the directions of the hybrids, has several important applications. Consider an atom, such as carbon, with four electrons in its valence shell, and able to use only s and p AOs to form bonds. If all four bonds are equivalent, each will have three times as much

† The chemical consequences of the structure of the water molecule are discussed by G. Pass in *Ions in solution* (3): *Inorganic properties* (OCS 7).

p-character as s-character. Each hybrid $p + \lambda s$ could therefore be written $s^{\frac{1}{2}}p^{\frac{1}{2}}$. Alternatively we can say that each hybrid is of the form $p + s/\sqrt{3}$, so that $\lambda = 1/\sqrt{3}$. It now follows from eqn (19) that the angle between any two such hybrids satisfies the equation $\cos \theta = -\lambda^2 = -\frac{1}{3}$. Thus θ is the familiar tetrahedral angle $109°\ 28'$, and we see that a carbon atom is indeed able to form four equivalent bonds, as in CH_4, and that these bonds point in the tetrahedral directions. Each C—H bond in methane will be described by two electrons with opposed spins in a wavefunction built from one of the tetrahedral hybrids of carbon and the 1s orbital of hydrogen.

The valence state

Methane can be regarded as the prototype of all compounds involving saturated carbon atoms. But our description has implied more than we may perhaps have realized. For in Chapter 1 we saw that the ground state of carbon has electronic structure $(1s)^2(2s)^2(2p_x)(2p_y)$. According to the earlier part of this chapter is should therefore be divalent (as it is in some unusual molecules such as CF_2; methylene CH_2 does exist, and plays an important role in many chemical reactions, but it is highly reactive, and cannot conceivably be regarded as normal for bonds involving carbon). In order to get four bonds around a carbon atom we need four unpaired electrons. So we must excite one of the pair of 2s electrons into the vacant 2p orbital, to give the arrangement $(1s)^2(2s)(2p_x)\cdot(2p_y)(2p_z)$. It is known from spectroscopic studies that if the spins of all four valence electrons are parallel, the resulting ^5S state lies about 436 kJ mol^{-1} above the ground state ^3P. Thus, in order to make carbon tetravalent an amount of energy of this order has to be put into it. Our reward for doing so is that we form four bonds instead of two. Now a C—H bond in methane has bond energy about 450 kJ mol^{-1}. It is therefore well worth our while to feed in this promotion energy and create a carbon atom in an sp^3 *valence state* since we recover about twice as much energy through the extra bonds thus made possible.

The valence state just described does not exist for an isolated atom. This is because, if we ever find an isolated carbon atom, it will be in one or other of its spectroscopic states, and not in a state involving hybrids of s and p. Furthermore the actual energy of the valence state plays effectively no chemical role, since it remains almost constant provided that the carbon atom is saturated.

Our description of the tetrahedral carbon atom depended on the fact that the excitation s → p needed only a relatively modest amount of energy. In understanding valence behaviour, therefore, it is necessary to know how the s–p difference varies from atom to atom. There are no simple rules,† but in

† See the discussion by R. J. Puddephat in Chapter 4 of *The periodic table of the elements* (OCS 3), and also that by G. Pass in *Ions in solution* (3): *Inorganic properties* (OCS 7).

general: (i) the s–p difference does not vary greatly among the members of any group in the periodic table, though it is usually rather larger in the first complete row (Li to Ne) than in later rows; (ii) the s–p difference increases rapidly along any row of the periodic table. Thus in oxygen it is twice as large as in carbon, and in fluorine approximately three times. It is for this reason that, as we shall see in the next chapter, atoms of group IV show the greatest versatility in bonding, with groups III and V next in interest.

Double and triple bonds in carbon compounds

The versatility of carbon, typical of group IV atoms, is shown by its ability to form double or triple bonds. Ethene will be a good starting point. The molecule is planar, and the valence angles are close to 120°. In benzene, which is also planar, the valence angles are exactly 120°; so let us discuss the carbon atom in terms of three bonds in a plane at 120° angles. The fundamental formula (19) now shows that the hybrids in the plane are of the form $p + \lambda s$, where $\lambda^2 = -\cos 120° = \frac{1}{2}$. Each hybrid is of the type $s^{\frac{1}{3}} p^{\frac{2}{3}}$, and collectively they arise from sp^2 in the atom. These three hybrids are shown diagrammatically in Fig. 27. They are usually referred to as trigonal hybrids. If the plane containing the axes of the three hybrids is the xy plane, then there remains the $2p_z$ AO of carbon, not used in forming the three hybrids.

The electron description of ethene is now quite straightforward. We imagine the two carbon atoms in their trigonal valence states, and form four C—H bonds by using in each case one of the hybrids of carbon and the usual 1s AO of hydrogen. Similarly a C—C bond is formed from the two remaining hybrids. The criterion of maximum overlapping requires that all the valence angles be 120°. But at this stage we have simply a set of five σ-bonds, and there is nothing to prevent almost free rotation around the central C—C bond. An electron count shows that there are two electrons not yet assigned to any orbitals in the molecule. If they are in the $2p_z$ AOs of the carbon atom, then, just as in Fig. 12(d), they can form a π-bond. But the need for maximum

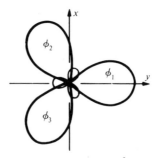

FIG. 27. The three trigonal sp^2 hybrids.

overlapping then requires that the two z-directions are indeed parallel: the molecular is planar. And if one CH_2 group is rotated relative to the other around the carbon–carbon bond, the overlap of the two p_z orbitals is reduced, and bonding is lost. Thus the valence angles of a molecule such as ethylene are determined primarily by the hybridization in the σ-bonds; but the resistance to torsion around the double bond is due to the π-bond and, as we might have expected, the double bond is represented by the superposition of a σ and a π bonding pair of electrons.

The fact that the valence angle H—C—H in ethene is not quite 120° (it is actually about 117°) may be attributed to the difference in overlap integral between two carbon hybrids and between a carbon hybrid and a hydrogen 1s orbital. The greater carbon–carbon overlap, increased as it is by the bond shortening due to the double bond, favours a greater s-character in the C—C bond. This implies a greater p-character in the C—H bonds and so, from eqn (19), a somewhat reduced H—C—H angle.

In ethyne (acetylene) H—C≡C—H we form digonal hybrids (Fig. 28) $s + p_z$ at each carbon atom, leaving the p_x and p_y orbitals unchanged. So we get three σ-bonds, all collinear; and the remaining electrons produce two π-bonds, just as in N_2 (p. 35). The molecule is linear, and the triple bond is a superposition of one σ- and two π-bonds.

Bent bonds: strain

In all the compounds just discussed it was possible to have straight bonds. This was because there were no steric limitations to prevent us placing each atom in such a way that maximum overlap was achieved with its neighbours. But sometimes there are steric limitations, and we speak of a strained molecule.† With our present model of a bond as due to overlapping between orbitals of the two atoms involved in the bond, we can soon see how this strain arises. The most important example (Fig. 29) is cyclopropane, which

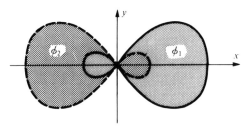

FIG. 28. The two digonal sp hybrids.

† See, for example, D. Whittaker's *Stereochemistry and mechanism* (OCS 5), Chapter 4.

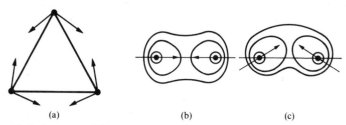

(a)   (b)   (c)

FIG. 29. Cyclopropane C_3H_6. (a) The equilateral triangle formed by the three carbon atoms, and the directions of the hybrids used to form localized C—C bonds. (b) Contours of charge density for a normal straight C—C bond (schematic). (c) Contours of charge density for a bent C—C bond (schematic).

consists of three carbon atoms at the vertices of an equilateral triangle, with the three pairs of H atoms located symmetrically above and below the plane of the triangle, such that the H—C—H angle is approximately 115°. This is distinctly larger than the normal tetrahedral angle of 109° 28′.

It is clear that if we construct four equivalent tetrahedral hybrids at each carbon atom, the pair of hybrids used for each C—C bond will not point directly towards each other, and the overlap will be reduced. If indeed we choose the hybrids such that two of them make an angle less than the tetrahedral one, they will overlap a little more effectively for the C—C bonds. Moreover in order to preserve orthogonality the other two hybrids will make a bigger angle than the tetrahedral one. Now the fundamental equation (19) shows that by no real combination of s and p is it possible to obtain a valence angle less than 90°. Some compromise has to be reached. It appears that this occurs when the hybrids for the C—C bonds make a mutual angle of about 100°, and then the C—H bonds, which can remain straight, will make an angle of about 115°, just as found experimentally. The directions of the arrows in Fig. 29(a) show the directions in which the in-plane hybrids point. As a result the bonds are bent (Fig. 29(c)), and their charge density differs from that of a straight bond (Fig. 29(b)).

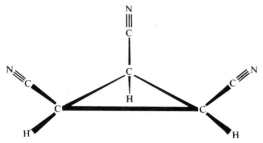

FIG. 30. The tri-cyano derivative of cyclopropane.

A most interesting verification of the existence of these bent bonds has been obtained by Hartman and Hirshfeld,[†] who studied the tri-cyano derivative in Fig. 30. The replacement of three hydrogen atoms by cyano groups will be expected to make very little difference to the electronic charge density in the C_3 plane. In this plane we have three bent C—C bonds. The overlap regions (p. 25) for the hybrids shown in Fig. 29(a) will not lie symmetrically along the lines joining the carbon atoms, but outside the triangle formed by them. Thus whereas the difference-density diagram for straight bonds should show (as it does in those cases that have been studied) a build-up of charge along the line joining the atoms, in the case of bent bonds it should occur off this line. This is exactly what is found, as Fig. 31 nicely illustrates.

Thus Baeyer's theory of strain in a molecule has its modern interpretation in terms of an inability to achieve maximum overlap through steric restriction; this leads to a decrease in the overlap binding energy, and hence to a smaller total binding energy (or heat of formation).

Fig. 31. Difference-density diagram for tricyanocyclopropane in the C_3-plane. (A. Hartman, personal communication.) The important overlap regions have been emphasized by shading.

Advantages and disadvantages of hybridization

It may be useful, before concluding this chapter, to review the importance of hybridization in terms of its advantages and disadvantages.

† Hartman, A. *and* Hirshfeld, F. L. (1966). *Acta Crystallogr.* **20**, 80.

The most important single reason for using the concept is that it allows us to continue to think of a chemical bond in terms of a two-electron function built up from particular orbitals ϕ_a, ϕ_b on the two centres. If we insist that ϕ_a and ϕ_b are simply pure s- or p-AOs, then, as the example of methane makes abundantly clear, there is no satisfactory way of preserving the simple picture of an electron-pair bond, originally developed for H_2 but fitting so well with the empirical conclusions reached by G. N. Lewis. For if the orbitals of the carbon atom in CH_4 are not hybridized, then one of the C—H bonds (the one which uses C(2s)) will necessarily be different from the other three (which use $C(2p_x)$, $C(2p_y)$, and $C(2p_z)$). This is completely contrary to experience.

A second advantage in the use of hybrids is that they give a very direct picture of enhanced bonding through increased overlap. The way in which the overlap depends on the mixing of s and p is shown in Fig. 32,† which gives the overlap integral S for two similar hybrids of carbon at a fixed distance apart. The largest value of S occurs for the digonal sp hybrids of Fig. 28; and it is very much larger than when using either pure-s or pure-p orbitals.

A third advantage arises incidentally as a kind of bonus. Let us refer once more to the sp^2 trigonal hybrids shown in Fig. 27, and consider the interaction between two electrons in distinct hybrids. The important part of

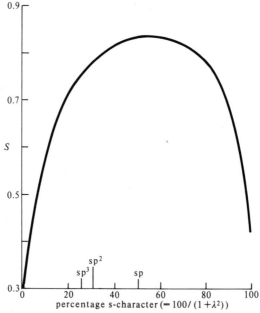

FIG. 32. The overlap integral S for two similar hybrids of the form $s + \lambda p$ for a carbon atom, as a function of the mixing parameter λ and at fixed internuclear distance.

† Taken from Maccoll, A. (1950). *Trans Faraday Soc.* **46**, 369.

the diagram is reproduced in Fig. 33(a). The major interaction is a simple Coulomb repulsion similar to the energy $e_1 e_2 / 4\pi\varepsilon_0 r$ between charges e_1, e_2 at a distance r apart. Now the mean positions of the electrons in the two hybrid charge-clouds lie at P and Q, well away from the nucleus. Moreover the electrons seldom get close to each other. Hence the Coulomb repulsion between the charge-clouds is smaller than if, as in Fig. 33(b), we consider an s- and p-pair of orbitals: this diagram shows that there is then a substantial probability that the electrons get close together, and so increase the mean value of $1/r_{12}$. The further P and Q lie away from the nucleus (i.e. the more strongly directional the hybrids) the smaller will be the Coulomb repulsion. So also will be the 'exchange effects' which must be taken into account whenever there is a substantial overlap region between two orbitals.† In this way we get a very clear picture of another of the energetic factors which determine the shape of a molecule; the total energy is lowered when the electron pairs are concentrated between pairs of nuclei (giving strong bonds) but it is also lowered *when the individual pairs keep as far apart as possible*, thus reducing the Coulomb repulsion among the electrons. This observation is the basis of the 'valence-shell-electron-pair-repulsion' (VSEPR) principle‡ much used in inorganic chemistry. If we try to build our wavefunctions from orbitals like those in Fig. 33(b), without hybridization, we obscure the simple physical picture.

A fourth advantage in the use of hybrids is that they allow us to understand the correct equivalence between similar bonds. The example of CH_4 is helpful, because quite clearly all four C—H bonds can be equivalent only if we use equivalent orbitals of the carbon atom. If these are hybrids, then, as we have seen, this is perfectly possible: otherwise it is not.

A final advantage is that the use of hybrids leads us to the correct valence angles. Thus we can see at once why, in H_2O, the valence angle is greater than $90°$, and why CH_4 is tetrahedral and not square planar.

(a)   (b)

FIG. 33. Reduced Coulomb repulsion between two electrons in trigonal hybrids (a), as compared with the repulsion between an s- and a p-orbital (b).

† See *CV*, Section 7.8.

‡ For a detailed account see R. J. Gillespie, *Molecular geometry* (Van Nostrand, Princeton 1972).

A difficulty with hybrids is that, unless the particular mixing of s, p, d, . . . is determined by symmetry alone (as it is often is), a further calculation of some sort is needed to determine the mixing parameters. Indeed, as we shall see in Chapter 5, there are situations in which no amount of hybridization will yield a simple picture of the bonding. Another objection to their use is that they are mathematically unecessary; provided we do an actual calculation, including everything in the mathematics, it does not matter whether we start from 'raw' AOs or first combine them together into hybrids. Against such objections, however, it may be argued that the use of hybrids often gives immediate physical insight into the nature of the bonds, and their mutual orientations, *even without calculations of any kind.*

One final comment—people sometimes talk of hybridization as if it were a phenomenon, something that actually happens. This is not so. It is perfectly possible to write down admirable wavefunctions for a polyatomic molecule, without any reference whatever to hybrids. The value of hybridization is conceptual, as we have stressed above; it extends the idea of a two-electron bond, with a localized charge cloud, from the case of simple diatomics such as H_2 and F_2 to polyatomic molecules; and without it we should find it hard to 'explain' the constancy of bond properties so characteristic of such molecules.

Different types of hybrid

There is no reason why our hybrids should consist solely of atomic s- and p-orbitals. A wide variety of combinations of different numbers of s-, p-, and d-orbitals exists, the details of which can often be most easily determined by group theory. Chemically, however, the condition for effective hybridization of two or more AOs is that they have roughly the same energy (ionization potential). An equivalent condition is that they have approximately the same size. It is this condition that normally excludes hybrids involving inner- and outer-shell orbitals.

In the next chapter we shall be concerned with valence angles in each group of the periodic table. So, because we shall then find it useful, we show in Table 2 a list of the more common types of hybrid, together with the directions in which these hybrids point.

One or two examples will serve to show how such hybrids may be constructed by superposition of the various types of s, p, d orbital described in Chapter 1. Thus, linear dp hybridization arises when we combine d_{z^2} and p_z AOs, 'adding' and 'subtracting' to get mixtures which point along the positive and negative z-axes, respectively, as in Fig. 34. On the other hand, octahedral d^2sp^3 hybridization occurs when we mix *two* d-orbitals, *one* s-orbital and *three* p-orbitals, (e.g. $3d_{x^2-y^2}$, $3d_{z^2}$, $4s$, $4p_x$, $4p_y$, $4p_z$) to provide six hybrids which point outwards (in the positive and negative directions) along the three coordinate axes—thus defining the octahedron in Fig. 59.

The same considerations apply to the hybrids in Table 2 as to those discussed in the last section. They allow us to extend the electron-pair concept

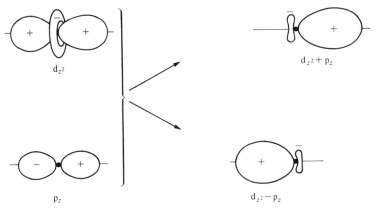

FIG. 34. Hybrids formed from d_{z^2} and p_z AOs.

to molecules involving polyvalent atoms, in which a single atom may be covalently bonded to as many as six neighbours. The main criterion for effective combination of AOs remains one of roughly similar extension in space, thus 2s and 4p would not be expected to combine appreciably, but 4p and 3d, provided they were of roughly similar size, might do so quite effectively; such intershell mixing occurs only in shells of principal quantum numbers 3, 4, 5, where all orbitals are comparatively diffuse.

TABLE 2

Important types of hybridization

Coordination number for hybrids	Atomic orbitals used	Resulting hybrids
2	sp	Linear
	dp	Linear
	sd	Bent
3	sp^2	Trigonal plane
	dp^2	Trigonal plane
	d^{2}s	Trigonal plane
	d^{2}p	Trigonal pyramid
4	sp^3	Tetrahedral
	d^{3}s	Tetrahedral
	dsp^2	Tetragonal plane
5	dsp^3	Bipyramid
	d^{3}sp	Bipyramid
	d^{4}s	Tetragonal pyramid
6	d^{2}sp^3	Octahedral
	d^{4}sp	Trigonal prism

Finally, it must be emphasized that the use of hybrids is not confined to polatomics. We saw on p. 46 that the bonds in diatomic Li_2 involved a small amount of hybridization. Furthermore, when listing the MOs for diatomics (Fig. 16) we implied that $2\sigma_g$ arose solely from the 2s AOs of the two bonded atoms, and $3\sigma_g$ from the $2p_z$ AOs. This is a simplification, since these two MOs, being of the same symmetry, can mix together as suggested by the Rayleigh-Ritz theorem, thus leading to MOs formed from hybrids of s and p. It is partly due to the fact that the extent of this hybridization varies from atom to atom that sometimes the sequence of energies of the $1\pi_u$ and $3\sigma_g$ MOs is reversed.

PROBLEMS

3.1. Show that instead of the form given in the text, the four tetrahedral hybrids of a carbon atom can be put in the alternative form $\frac{1}{2}(s \pm p_x \pm p_y \pm p_z)$, where s, p_x, p_y, p_z are normalized AOs, and we choose either one or three of the positive signs.

3.2. Verify that sp^2 trigonal hybrids may be written in the normalized form $\sqrt{\frac{1}{3}}(s + \sqrt{2}p_x)$, $\sqrt{\frac{1}{6}}(\sqrt{2}s - p_x \pm \sqrt{3}p_y)$. Hint: show that these hybrids are normalized, and orthogonal; and that they point in directions in the xy-plane that make angles of $120°$ to one another.

3.3. Use the fundamental eqn (19) to show that hybrids with equal amounts of s and p_z are collinear.

3.4. What would you expect to happen to the valence angle in H_2O if the s–p energy difference in oxygen were reduced to zero?

3.5. Why is CO_2 linear and NO_2 bent?

3.6. In H_2O (p. 47) we claimed that the AOs to be used by O are hybrids of s and p. Why is the same concept of hybridization not important for the H orbitals also?

3.7. P_4 is a tetrahedral molecule. Where would you expect the difference-density diagram to show a build-up of charge? Can you make any suggestions why P_4 is stable, but N_4 is not?

4. Valence rules

Valence rules—preliminary statement

WE have now completed our survey of the methods used to discuss chemical bonding, and are ready to apply them to different species of atoms. The Mendeleev periodic table† shows us that it is sensible to think of all atoms as being arranged in groups. The front cover shows the conventional modern arrangement, and the group number of each column. The rare earths and the trans-uranic elements (lanthanides and actinides) shown separately at the bottom of the table will not concern us in this book, since they represent the filling up of f-AOs. The three transition-element series, to which group numbers are not always assigned (though they are often referred to as the A-series to distinguish them from the B-series at the right hand end of each long row of the table), similarly represent the filling up of d-AOs. As we saw in Table 2, there are interesting types of hybrids which result from suitable combinations of s-, p-, and d-orbitals.

Our picture of a normal single bond (σ-bond) is based on two electrons, usually one from each atom, with opposed spins and approximately described by a wavefunction of VB or localized-MO type. This wavefunction is built up from two orbitals, associated with the two atoms, and chosen to overlap as strongly as possible. In polyatomic molecules, and in diatomics too, these orbitals will be appropriate hybrids; and the hybrids around any one central atom must be orthogonal. Such hybrids are energetically 'profitable' only if their component s-, p-, and d-orbitals have similar energies. The hybrids for the two atoms of a bond will normally point directly towards each other, leading to a straight bond: if steric considerations prevent this, the bonds will be bent, and somewhat reduced in strength. A double bond is a combination of a σ- and a π-bond; a triple bond is $\sigma\pi^2$. In all this it makes relatively little difference whether we are using the VB model or the localized-MO model. What is important is that the valence shell of an atom should contain singly occupied AOs (or hybrids); the normal valence number of the atom will then be the number of such orbitals—in either the ground state or some readily accessible valence state.

Sometimes even when the valence number is apparently zero, all valence AOs being filled, a covalent bond can be formed—but only by 'donation'. Thus, if a doubly occuped AO of an 'electron-rich' atom is combined with an empty AO of an 'electron-poor' atom, the two electrons available may still fill the resultant bonding MO to give a *dative* or *donor* bond. We shall find that

---

† The structure of the table and its role in chemistry are described by R. J. Puddephatt in *The periodic table of the elements* (OCS 3).

atoms with lone pairs, in groups such as NH_3 or CO, often act as donors while transition-metals, with incompletely filled d-shells, often act as acceptors.

With these preliminary observations we are ready to discuss valence behaviour, group by group. Let us begin with group I.

Group I: the alkali atoms

The characteristic of all group I atoms is that they have one valence electron outside an inner closed-shell structure. In the case of lithium this is a 2s electron; with sodium it is a 3s, and so on. With one unpaired valence electron we expect monovalence, and this is just what we find. Examples are Li_2 and LiF. The outer electron is a very diffuse one, so that its ability to overlap with an orbital from some other atom is small. Consequently covalent bonds involving these atoms are weak and long. Stronger bonds occur in polar molecules, where the group I atom has largely given up its valence electron, and, then being much smaller in size, can approach more closely to the ligand. This situation is revealed by the very large dipole moments of such diatomic molecules. Thus for the diatomic alkali halides CsF, KBr, and NaCl the observed dipole moments are 7.9, 9.1, and 8.5 debye. These may be compared with water (1.8 debye) and ammonia (1.5 debye).

Since the s-type valence electron occupies a large-size orbital with no directional character it can overlap reasonably well with a number of neighbours simultaneously located around it. Partly for this reason these atoms have a strong tendency to form metals, in which the bonds are completely delocalized (Chapter 5).

When a group I atom forms a bond, as in Li_2, it will add a certain amount of p-orbital to mix with the original s-orbital, in order to get better overlapping. The percentage of p-character for diatomic molecules of this group has been estimated by Pauling[†] to be as follows (Table 3).

These relatively small values show that the bonds are sensibly described as s-bonds. Experimental measurements of the charge density have not been made, but Fig. 35 shows the theoretical contours for Li_2. The inner-shell pairs $(1s)^2$ around each nucleus are clearly revealed. The outer part corresponds to the bonding pair, and, as would be expected, is similar in shape to that shown in Fig. 10 for H_2, where the bonding comes from the 1s orbital rather than the 2s, as for lithium. In MO language the bonding pair (which alone contributes to the difference-density diagram in Fig. 21) is in the $(2\sigma_g)^2$ molecular orbital. If we remove one of these electrons, to get Li_2^+, the bonding power of the one remaining $2\sigma_g$ electron is sufficient to keep the molecule intact. Indeed the mutual repulsion between the $2\sigma_g$ pair in Li_2 is about equal to the bonding power of each, so that after removing one of them, the dissociation energy is not much altered. In fact the value for Li_2^+ is 1.55 eV, slightly greater than that for Li_2 (1.12 eV). This indicates very clearly how weak these bonds are.

[†] Pauling, L. (1949). *Proc. R. Soc.* A **196**, 343.

TABLE 3

Hybridization in alkali-metal diatomic molecules

Molecule	Li_2	Na_2	K_2	Rb_2	Cs_2
Percentage p-character	14.0	6.8	5.5	5.0	5.5

Group VII atoms: the halogens

Monovalent atoms also occur in group VII.† These are the halogens F, Cl, Br, and I, typified by having one unpaired electron in an otherwise complete shell. Thus F is $(1s)^2(2s)^2(2p)^5$, whereas the rare gas neon is $(1s)^2(2s)^2(2p)^6$. As there is just one unpaired electron the normal valence number is one, and the atoms are monovalent. The existence of HF, F_2, and ClF all show that this expectation is correct. Since (p. 50) the s–p energy difference for atoms near the end of any row of the periodic table is high, there is little hybridization, and the bonds are therefore almost completely p-bonds. Figure 22 shows that if, as in HF, there is a considerable difference in electronegativity of the atoms concerned, the bond is distinctly polar ($\mu_{HF} = 1.82$ debye, leading to a formal

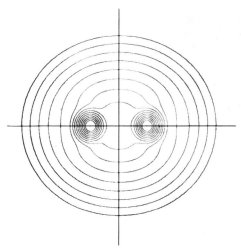

Fig. 35. Contour diagram for total electronic charge density in Li_2. (The corresponding difference density is shown in Fig. 22.) (*Science* **151**, 961 (1966) courtesy of A. C. Wahl.)

† More strictly, group VIIB.

charge distribution $H^{+0.4}F^{-0.4}$, and the almost complete absorption of the proton within the charge-cloud of the fluoride ion). We have already seen, in Table 1, how the decreasing electronegativity difference as we go down the appropriate column of the periodic table leads to a decreasing polarity in HCl, HBr, and HI.

At first it might be thought that this was all there was to say about the halogens. This is not so. There are most interesting higher valences. Thus we find stable ClF_3 and BrF_3. Moreover (Fig. 36) these molecules are T-shaped, with the 'vertical' bond shorter, and therefore presumably stronger, than the two 'horizontal' ones. There are also more complicated molecules such as BrF_5 which is approximately a tetragonal pyramid with the Br atom just below the plane of the four F atoms, and IF_7. The simplest way to understand these molecules is by modification of a model which we shall later find useful for understanding rare-gas compounds (p. 69). Let us illustrate it by considering ClF_3.

We begin (Fig. 37(a)) with a normal Cl—F molecule, described in the usual way by a σ-bond, and built out of $Cl(3p_z)$ and $F(2p_z)$, where the z-axis is taken along the Cl—F direction. Around the chlorine nucleus there are various orbitals, such as the $3p_x$ AO, not used in the Cl—F bond. If we ionize an electron from one of these we then have an unpaired electron with which (Fig. 37(b)) we can make a second Cl—F bond, at right angles to the first. The ionized electron must go somewhere: so let us bring up a third F atom to accept it, giving the bond structure in Fig. 37(b). It is obvious that we could have reversed the roles of the second and third fluorine atoms, getting Fig. 37(c). We now use the Rayleigh-Ritz principle (p. 15) and say that since we may expect the complete wavefunction to have properties characteristic of both diagrams, we could represent it in the form $\Phi = c_1\Phi_1 + c_2\Phi_2$, where Φ_1 and Φ_2 are wavefunctions corresponding to (b) and (c): and use the variation method to get c_1 and c_2. By symmetry, however, we may speak of resonance between the two valence structures Φ_1 and Φ_2.

This model leads to the second and third F atoms lying on the axis of the $3p_x$ chlorine orbital, and hence to the observed T-shape. It also suggests that the bonds in the arms of the T are less strong than normal single bonds, and so leads to an explanation of the differences in bond lengths shown in Fig. 36.

Other polyhalogenide systems can be understood in a similar way.

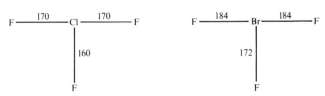

FIG. 36. The T-shaped molecules ClF_3 and BrF_3. Bond lengths are shown in pm.

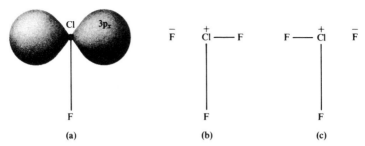

FIG. 37. Bonding in ClF_3. (a) The $3p_x$ orbital involved in the linear F—Cl—F bonds. (b), (c) Contributing bond structures.

Group VI atoms

We turn next to the group VI atoms, O, S, Se, Te, and Po. Here the outer electronic structure is s^2p^4. If we write this in the form $s^2p_xp_yp_z^2$ we see that there are two unpaired electrons. Such atoms should have a normal valence of two. These Group VI atoms are indeed most frequently divalent. We have already discussed the H_2O molecule (p. 41), and shown how, without hybridization, we should expect a 90° valence angle. However, if we hybridize the s and p AOs we expect a somewhat larger angle. The observed values for the dihydrides are shown in Table 4. There are two reasons why the valence angle in H_2O exceeds 90°. One is the use of hybrids of s and p previously described; the other is purely electrostatic. For since O is more electronegative than H, we expect each O—H bond to involve covalent–ionic resonance (p. 29), leading to a resultant net positive charge around each proton. These two positive charges will repel each other with a Coulomb inverse-square-law force, and so the angle H—O—H is increased. Reasonable assumptions about the charges suggest that perhaps 5° of the increase beyond 90° may be attributed to this; the rest may be associated with hybridization. Since the electronegativity difference decreases as we go down the group, we expect the electrostatic part of the increase in valence angle to decrease as well. But there is as yet no satisfactory simple explanation of the fact that, except for H_2O, all the valence angles are close to 90°, implying p-bonds with little or no hybridization.

If the oxygen atom in H_2O is ionized, we have one more unpaired electron, so that a third bond can be formed. In this way we understand the existence of

TABLE 4

Valence angles in Group VIB hydride molecules

Molecule	H_2O	SH_2	SeH_2	TeH_2
Valence angle	$104\frac{1}{2}°$	93°	91°	$89\frac{1}{2}°$

oxonium compounds, such as H_3O^+, which are expected to be pyramidal and not planar.

Once again higher valences exist. Thus there are $TeCl_4$ and SF_6. Let us consider SF_6. This (like TeF_6) is an octahedral molecule. If we want to describe the bonding in terms of electron pairs, we need to create six octahedral hybrids around the S nucleus. A glance at Table 2 shows that this may be achieved with a combination sp^3d^2. Since the normal ground state of S is s^2p^4, the creation of the appropriate valence state would require us to promote one s- and one p-electron into empty d-orbitals. The energy of such a promotion is rather high. However, by doing so, we gain no less than four new bonds. It must therefore be considered a possibility. The energetics of such a process are not yet fully understood. Part of the difficulty lies in the fact that for an isolated S atom the 3d orbitals lie well outside the 3s and 3p orbitals, so that, unless some mechanism exists for compressing the d-orbitals, and not the s- and p-orbitals, effective hybridization cannot be achieved. One such mechanism is known: if the S atom is partly ionized (as it would be with polar S—F bonds) then it appears that the d-orbitals contract. The matter is still not settled, and an alternative description, not involving any d-orbitals, is discussed in one of the problems at the end of this chapter.

Group V atoms

The group V atoms N, P, As, Sb, and Bi form bonds in a manner very similar to that of the group VI atoms. There are three unpaired electrons in their valence shells. So the atoms are normally trivalent. The unpaired electrons are in p-orbitals, so that, without hybridization, we expect valence angles of around 90°. Table 5 shows that this is just what is found experimentally, except for ammonia NH_3. Just as with group VI this shows that except for NH_3 the bonds use almost pure p-orbitals. In ammonia, however, the valence angle is only a little less than the tetrahedral angle $109\frac{1}{2}°$, so that we must suppose that the N atom is hybridized almost tetrahedrally. This would imply that the two lone-pair electrons have a charge-cloud projecting away from the N nucleus in much the same way as was illustrated for H_2O in Fig. 26. It is not surprising, therefore, that ammonia very easily adds a proton to form tetrahedral ammonium NH_4^+. In this process the N—H bond length changes only from 101 pm to 103 pm. This shows how well localized each N—H bond must be.

TABLE 5

Valence angles in Group VB hydride molecules

Molecule	NH_3	PH_3	AsH_3	SbH_3
Valence angle	107°	$93\frac{1}{2}°$	92°	91°

Electrons in lone-pair orbitals such as those of ammonia exert strong repulsion on any other similar electrons. This repulsion is of great importance in determining the shape of hydrazine $H_2N—NH_2$. Around each N the electronic distribution resembles that of ammonia. But if (Fig. 38) we place the two NH_2 groups in an eclipsed configuration, the two sets of lone pairs are close to each other, and there is a strong repulsion. As a result the NH_2 groups are rotated around the N—N bond away from the eclipsed configuration. Experimentally the azimuthal angle is about 95°.

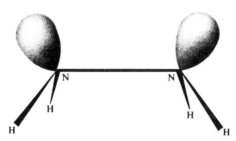

FIG. 38. Hydrazine N_2H_4. The diagram shows the unstable eclipsed configuration in which the lone pairs, shown shaded, exert a large repulsion on each other, leading to rotation of one NH_2 group relative to the other, around the N—N axis.

There are many other diverse compounds of this group. We have space to discuss just one, the case of penta-coordination. Familiar examples are PCl_5, PF_5, AsF_5 and $SbCl_5$. All are trigonal bipyramids, as shown in Fig. 39 for PCl_5. Here the equatorial P—Cl bonds, marked eq, are stronger and shorter (204 pm) than the axial bonds ax (219 pm). Now Table 2 shows that five hybrids in trigonal bipyramidal form can be constructed from sp^3d. Since the ground state of P is s^2p^3 this would involve a promotion $s \rightarrow d$ in order to create the appropriate valence state. Such a promotion gives two additional bonds; it can become possible, just as in SF_5 discussed earlier, if the 3d orbital can be contracted. However, there is an alternative scheme, not needing d-orbitals. We first use phosphorus sp_xp_y to form trigonal hybrids (p. 50) and so complete the equatorial P—Cl bonds. There would then remain an unused pair $P(3p_z)^2$ where the z-direction is along the vertical axis of the diagram. We could now use this pair just as we used a similar pair in ClF_3 (see Fig. 37) to bond two more electronegative atoms along this axis. Such a model does not require the use of any d-orbitals, and it correctly predicts that the equatorial bonds should be stronger than the axial ones. In favour of the use of d-orbitals it is often argued that these penta-coordinated molecules are not found when the central atom is nitrogen: and of course there are no d-orbitals in the

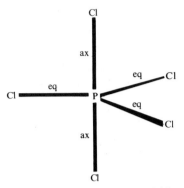

F<small>IG</small>. 39. The trigonal-bipyramid molecule PCl$_5$. The axial bonds ax are longer and weaker than the equatorial bonds eq.

valence shell of nitrogen (d-orbitals begin with 3d) whereas there are fairly accessible d-orbitals for all the heavier atoms of this group.

Group II atoms

Group II atoms (Be, Mg, Ca, $\cdots$) have two electrons in their valence shell. But since in the ground state these electrons are paired together in the form $(ns)^2$, where $n = 2$ for Be, 3 for Mg etc., we should conclude that the natural valence number for each of these atoms is zero. However, promotion to a valence state based on sp leads to two unpaired electrons, and hence divalence. Such an excitation is not too 'expensive', and provides two additional bonds; nor is this the only possibility, as we shall see in a moment.

Let us consider a few of the many divalent compounds, beginning with the oxide BeO. At first sight, the Be atom would appear to be in a valence state with configuration $(1s)^2(2s)(2p)$; the 2s electron would participate in a σ-bond and the 2p electron (the 2p AO lying perpendicular to the bond) in a π-bond—thus giving a double bond with the oxygen. Spectroscopic evidence, however, points to a ground state with electron configuration BeO$[(1\sigma)^2(2\sigma)^2(3\sigma)^2(4\sigma)^2(1\pi)^4]$ and this assignment is supported by *ab initio* MO calculations. The $(1\pi)^4$ indicates that *both* Be valence electrons are involved in π-bonding, and therefore suggests a valence configuration Be$[(1s)^2(2p)^2]$ with an empty 2s AO; if the oxygen lone pairs are regarded as occupying digonal sp hybrids, one of these can then nicely overlap with the empty Be 2s to provide a *dative* bond (3σ) while the other (4σ) remains a lone pair. The two lowest-energy MOs (1σ and 2σ) are essentially atomic and describe the atomic K shells. The bonding in diatomic group II compounds is certainly not simple, as this example shows; but in other cases more familiar valence states appear to be appropriate.

The case of mercury dimethyl $HgMe_2$ is typical of many other compounds such as MgF_2, $MgCl_2$, and $MgBr_2$, all of which are linear. The ground state of Hg is $(6s)^2$, and promotion to the valence state 6s6p should lead to divalence. However, if we are to make two equivalent bonds we cannot use 6s and 6p as they stand: we must use the two digonal hybrids $6s \pm 6p$. Figure 28 now shows that the two σ-bonds thus made possible lie in diametrically opposite directions, this suggesting a linear molecule.

Some very interesting confirmation of this picture of $HgMe_2$ is obtained calorimetrically. The dissociation energy needed to break the first Hg—Me bond is 213 kJ mol^{-1}. but that needed to break the second such bond is only 23 kJ mol^{-1}. The explanation of this apparent lack of constant bond property is that when we take away the second methyl group, the mercury atom is left alone, and will therefore revert from an sp state to its ground state s^2. In the latter process we recover the valence-state energy (perhaps modified a little by the absence of the first methyl group). Thus, even if it requires the same energy to break the second Hg—Me bond (without changing the hybridization) as to break the first, this recovery of promotion energy would make it appear much smaller, as indeed it does. Without this recovery it would be hard to understand the great difference between the first and second dissociation energies. Similar results are found for other molecules, such as CH_3—Hg—X and X—Hg—X, where X is Br, Cl, or I.

There is a very amusing difference between elements in the IIA and IIB categories. Hg belongs to the IIB group, Ba to the IIA. The difference between these two atoms is that, although both have outer-shell electrons $(6s)^2$, in Ba the 5d shell is empty, whereas in Hg it is full. Consequently the lowest excitation in Hg is from 6s to 6p, but in Ba it is from 6s to 5d. So the two valence states will differ. We have already seen that in Hg the hybrids $s \pm p$ lead to bonds at an angle of 180°. In Ba, however, Fig. 40 (and Table 2) shows that the hybrids $s \pm d$ lead to bonds at an angle of 90°. Not very much is known about ths shapes of barium compounds, but it is interesting that BaF_2 is angular, and not linear, just as would be expected from this picture.

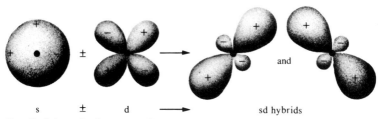

s   ±   d   ⟶   sd hybrids

FIG. 40. Schematic diagram to illustrate that sd hybridization leads to an angular molecule.

Group III atoms

Group III atoms are easy to deal with, at this stage. The valence shell of boron is $(2s)^2(2p)$, so that the natural valence number is one. Monovalence is found in BH, BF, BBr, and BCl. But more frequently we promote the atom to a valence state based on $2s2p_x2p_y$: then there will be three bonds. If all three are equivalent, as in BF_3 and BCl_3, we shall use sp_xp_y to form trigonal hybrids, as in Fig. 28. Then we expect a planar molecule with a threefold axis of symmetry. This is precisely what is found.

However, as soon as a fourth orbital is used the atom ceases to be trigonal. Thus just as NH_3 can add a proton to give tetrahedral ammonium NH_4^+, so BF_3 can add a fluoride ion to give tetrahedral BF_4^-. So also we find BH_4^- in crystalline $NaBH_4$, KBH_4, and $RbBH_4$.

The tendency of a boron atom to use its fourth valence-shell orbital and acquire a closed-shell octet of electrons is shown not only in the tetrahedral ions just described, but in the fact that, although BH_3 does not exist, diborane B_2H_6 does. Moreover B_2H_6 is only one of a large number of boron hydrides, with important chemical properties. We shall not describe the bonding in this molecule at this stage, but reserve it till the following chapter: this is because so far we have concentrated on localized bonds joining two atoms, and no satisfactory account of the boron hydride field can be given in terms of such bonds. We content outselves with a reference to the ammonia–borine molecule $H_3N \cdot BH_3$, where by electron donation to a state represented by $H_3N^+ - BH_3^-$ we are able to establish tetrahedral character at both ends of the central bond, and, in addition, find a reasonable nitrogen–boron bond length 156 pm. This is almost identical with the distance 154 pm to be expected by addition of the covalent radii 81 pm and 73 pm for B and N.

Group IV atoms

Group IV atoms add very little that is novel to the different possibilities illustrated for the earlier groups, except perhaps that almost all of them occur depending on circumstances. We have already seen (p. 49) how the normal valence number is two, but that a relatively low-energy promotion $s^2p^2 \to sp^3$ allows a valence state with valence four. If we use sp^3 to form four equivalent hybrids (p. 49) the familiar tetrahedral valence angles associated with a saturated atom (e.g. CH_4, SiH_4, GeH_4, C_2H_6) appear very naturally. If we use sp^2 to form trigonal hybrids (p. 50) we have three planar bonds at $120°$ angles, and the possibility of a π-bond, as in ethylene or formaldehyde $H_2C{=}O$. If we use sp to form digonal hybrids, as in Fig. 28, we have the collinear hybrids typical of an acetylenic carbon, as in $HC{\equiv}CH$. Almost pure p-bonding occurs in the radical CH, where there is little to be gained by destroying the s-character of the low-energy $(2s)^2$ group.

Nevertheless the great versatility of the group IV atoms, and particularly carbon, does permit us to study certain small differences, not so frequently or so fully available elsewhere. The best example is the C—H bond. Our previous paragraph shows that in forming this bond the carbon atom uses hybrid orbitals with varying amounts of s- and p-character. We require the overlap integrals for these hybrids (this time between C and H rather than, as in Fig. 32, between C and C: but the general result is just the same). It can be shown that the greatest overlap is with the sp digonal hybrid, followed by the sp^2 trigonal and then the sp^3 tetrahedral one. We therefore expect the acetylenic C—H to be a stronger and shorter bond than the ethylenic (or aromatic) C—H bond, and this itself should be stronger and shorter than the paraffinic C—H bond. This is precisely what is revealed in Table 6.

Even if these figures are slightly changed as a result of better experiments there is no doubt but that they completely confirm the expected type of behaviour. Similar results are found for C—Cl bonds, though here the matter is a little more complicated because there is a small amount of π-bonding as well as the pure σ-bonding of the C—H bonds. It should be noticed from the values in the table that the covalent radius of a carbon atom in various types of hybridization varies almost linearly with the amount of s- or p-character.

TABLE 6

Properties of C—H bonds involving different hybridization

Hybridization	Molecule	C—H bond length (pm)	Stretching force constant (N m^{-1})	Approximate bond energy (kJ mol^{-1})
sp	Acetylene	106.1	639.7	500
sp^2	Ethylene	108.6	612.6	440
sp^3	Methane	109.3	538.7	430
(p)	CH radical	112.0	449.0	330

The ideas just described for carbon must be quite general in their application. In this way we can understand that different covalent radii should be assigned to atoms, depending on the number and type of coordination involved.

The only common types of bonding involving carbon which have not been covered in the above discussion are those of conjugated systems such as butadiene and aromatic systems such as benzene. These cannot be described satisfactorily in terms of localized bonds. Their discussion, therefore, is reserved for the following chapter, under the general heading of many-centre bonds.

Group VIII atoms: the rare gases

Group VIII atoms, He, Ne, Ar, Kr, Xe, and Rn, sometimes called group O, have a closed valence shell: thus in He it is $(1s)^2$, but in all the others it is s^2p^6.

Since there are no unpaired electrons, we expect a normal valence number zero. This is often found, and is the justification for calling them the noble gases. However, this does not mean that they never combine to form molecules.

One way to bond a rare-gas atom is to ionize it, for then it has one unpaired electron and should resemble the halogen immediately preceding it in the table. So ar$^+$ resembles Cl (strictly, is *isoelectronic* with it) and we therefore expect ArH$^+$ to be stable, and isoelectronic with ClH. In fact its bond energy is 394 kJ mol^{-1}.

The case of He$_2^+$ is interesting. The molecule is quite stable ($D = 258$ kJ mol^{-1}) with a MO description $(1\sigma_g)^2(1\sigma_u)$. Two electrons are in bonding MOs, one in an antibonding MO, both MOs being built from the same set of 1s AOs; we call it a three-electron bond.

Another way to get bonding involving rare-gas atoms is to excite one of them. This leaves one electron in the original valence shell which now becomes available for bonding. Thus, although He$_2$ does not form a stable chemical bond, this molecule has a very rich electronic spectrum, all arising from states that dissociate into one (or two) excited He atoms.

The third kind of rare-gas bonding is much the most interesting, even though it was only discovered in 1961. Since ICl$_2^-$ is known to be stable, and linear; and since Xe is isoelectronic with I$^-$, we might expect that XeF$_2$ would be stable and linear. This is correct. So also is KrF$_2$. The explanation of this situation follows just the same general lines as on p. 62 where we were dealing with ClF$_3$. We start with an isolated Xe atom, and ionize it. We then let it bond with a F atom. Since it is known that $D(\text{XeF}^+) > 200$ kJ mol^{-1} this is quite an average sort of bond. The ionized electron is now placed on a second F atom (Fig. 41(a)). The resulting bond diagram will have its mirror image, shown in Fig. 41(b). Resonance between these leads to a stable linear molecule.

(a)   (b)

FIG. 41. The two chief resonating bond structures in XeF$_2$.

A large family of compounds of this sort has now been obtained. Since the starting-point involves ionization of the rare-gas atom, these molecules are stable if this ionization energy is not too large. It is for this reason that compounds of this kind are known involving Kr, Xe, and Rn, but not Ne, He, and Ar. Several oxides exist. Thus (see problems at the end of the chapter) XeO$_3$ is pyramidal, XeO$_4$ is tetrahedral, XeF$_4$ is square planar, and XeF$_6$ is approximately octahedral.

PROBLEMS

4.1. Show that if the carbonate ion CO_3^{2-} is written in the form $C^+(O^-)_3$, then a simple explanation of its plane trigonal shape is easily given in terms of appropriate hybrids of the carbon atom.

4.2. Why would you expect that ICl_2^- should be linear?

4.3. Why is it that in H_2O_2 we do not have a planar molecule, but the two OH directions correspond to a rotation of about 100° around the O—O axis from *cis*-planarity?

4.4. Use the ideas illustrated in Fig. 37 to provide a model of SF_6 that makes no use of d-electrons. Can you now see why SF_6 is stable, but SH_6 is not?

4.5. Use of the model of Fig. 41 to show that: (a) XeF_4 should be square planar, (b) XeO_3 should be pyramidal, and (c) XeO_4 should be tetrahedral. All these predictions are correct.

4.6. Describe the bonding in $TeCl_4$, which is known to have the structure shown. The equatorial valence angle θ is about 90°. Would you expect the axial or the equatorial Te—Cl bonds to be the stronger?

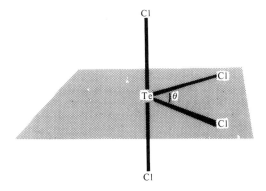

5. Delocalized bonds

Many-centre bonding

WE have already claimed that some types of bonding are not readily described in terms of localized two-electron two-centre bonds. These require delocalized orbitals, and unless we are prepared to extend our idea of a bond so that it becomes not just a two-centre but a many-centre concept we shall not easily understand these new situations.

Consider, as the simplest example, the H_3^+ molecule, which is known to be stable and to form an equilateral triangle (Fig. 42). There are only two electrons, but these two electrons must be able to hold all three atoms together. If we talk at all about one-electron orbitals, then this molecule is clearly best described by a two-electron three-centre bond. In the language of Chapter 2, instead of the two-centre MO $\phi_a + \phi_b$ we now have a three-centre MO $\phi_a + \phi_b + \phi_c$. Two electrons in this delocalized MO, with opposed spins, provide the necessary binding energy. This picture is much simpler than that obtained in the old style two-centre language. Then we should be obliged to speak of resonance between the three bond diagrams 42(b), (c), and (d).

This last example suggests another way of looking at a problem that we dealt with on p. 70—the molecule XeF_2. In Fig. 41 we represented this molecule in terms of resonance between two bond structures. These involved the Xe 5p orbital, directed along the molecular axis, and the two F 2p orbitals, whose positive directions are conveniently taken to be as shown by the arrows in Fig. 43. Following the technique used a little earlier for H_3^+ we use the LCAO approximation, and form MOs out of the three AOs F_a, F_b, and Xe. All three AOs have σ-symmetry; but there is also a centre of inversion, leading as on p. 33 to a u-, g-character in each MO. Since the Xe orbital is odd (u-character), it cannot appear in any g-type MOs. Thus the three delocalized MOs that we can build out of the three AOs are of the form

$$1\sigma_g = F_a + F_b + \lambda Xe,$$

$$1\sigma_g = F_a - F_b,$$

$$2\sigma_u = F_a + F_b + \mu Xe,$$

where λ and μ are two numerical parameters (of opposite sign because of orthogonality of $1\sigma_u$) to be determined by Rayleigh's variation principle. These three MOs are in order of increasing energy so that the four available electrons are fitted into the configuration $(1\sigma_u)^2(1\sigma_g)^2$. There is effectively no overlap between F_a and F_b, so that $1\sigma_g$ is nonbonding. Thus the bonds in XeF_2 are due to a two-electron three-centre orbital $(1\sigma_u)^2$. The reader will readily recognize that maximum overlap in $1\sigma_u$ is obtained when all three atoms are collinear, just as we found in our earlier resonance treatments.

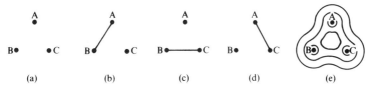

FIG. 42. H_3^+. (a) The equilateral-triangle equilibrium shape. (b), (c), (d) Component bond diagrams when using localized bonds. (e) Molecular orbital, containing two electrons, forming a two-electron, three-centre bond.

Diborane

This type of discussion enables us to give a simple description of diborane B_2H_6, which we were not able to deal with adequately in terms of two-centre bonds on p. 68. Figure 44(a) shows the molecule: the two BH_2 groups at each end lie in a single plane, and the bridge hydrogens H_1 and H_2 lie symmetrically above and below this plane. If we were to suppose that each line in the diagram represented a normal single bond we should find that we needed sixteen electrons, whereas there are only twelve available for us. But if we imagine that the external B—H bonds are of normal σ-type, with two localized electrons in each of them, this leaves us with four unused electrons. We can then fit these into two three-centre MOs, as shown in the diagram (Fig. 44(b)). In this way the correct number of electrons is used, and we see at once why the bridge hydrogens lie symmetrically above and below the plane of the other six atoms. Moreover, if the hybridization around each B atom is approximately tetrahedral we can also see how all four valence-shell orbitals s, p_x, p_y, p_z of each atom are used. In a generalized sense each boron atom contrives by this means to have its full octet of electrons.

Arguments of this kind can be used to describe all the many boron hydrides at present known. But the important point is that without delocalized MOs this would be far more difficult.

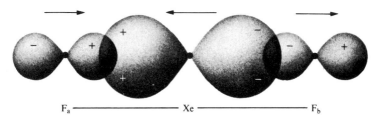

F_a ————————— Xe ————————— F_b

FIG. 43. Atomic orbitals used in the three-centre bonds in XeF_2.

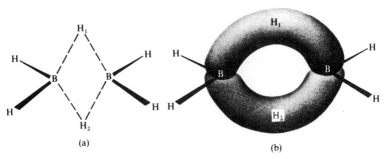

FIG. 44. Diborane B_2H_6. (a) The geometrical shape of the molecule. (b) The two three-centre two-electron bonds that bind the bridge protons H_1 and H_2.

Walsh diagrams

An interesting problem first studied in detail by A. D. Walsh† concerns the changes in shape that may occur when a molecule is ionized or electronically excited. Thus whereas the ground state of H_2O is triangular, some of its excited states and at least one state of its positive ion appear to be almost, or exactly, linear. Why is this? If we use the localized-bond description of p. 44 the answer is not obvious. Thus, if we think of one of the two O—H bonds as being ionized, we shall be wrong; since we cannot tell which of two identical bonds has lost an electron we should at very least have to consider an excited-state wavefunction involving both. But Walsh showed that if we use completely delocalized MOs the situation is much simpler and more instructive. The argument is as follows.‡

The water molecule (Fig. 45) is of C_{2v} symmetry. This means that the fully

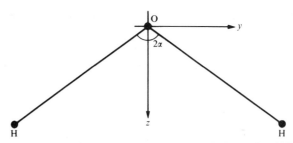

FIG. 45. The water molecule H_2O. The x-axis is perpendicular to the HOH plane.

† Walsh, A. D. (1953). *J. chem. Soc.* 2260–2331.

‡ A simple account is in a review by Coulson, C. A. (1970), Chapter 6 of *Physical Chemistry* Volume 5 (Valency) ed. by Eyring, Henderson, *and* Jost, Academic Press, New York and London, p. 315.

delocalized MOs belong to one of four possible symmetry classes: a_1, a_2, b_1, b_2. MOs of types a_1 and a_2 are unchanged by rotation of 180° around Oz, MOs of types b_1 and b_2 change their sign in this operation. Similarly MOs of types a_1 and b_2 are unchanged by reflection in the molecular plane, those of types a_2 and b_1 change in sign. This defines completely the symmetry properties of each molecular orbital.

There are eight valence-shell electrons; and they occupy the MOs $1a_1$, $1b_2$, $2a_1$, $1b_1$, where the sequence is that of increasing energy. Walsh next considered how the energies of these four MOs would change if the valence angle 2α were altered. He was led to the set of curves in Fig. 46.

According to this diagram an electron in the $1a_1$ MO has lower energy, i.e. better binding, if the valence angle increases. The same is true for an electron in $1b_2$. Precisely the opposite is found for $2a_1$, whereas an electron in $1b_1$ hardly affects the valence angle. In the ground state of the molecule all four MOs are doubly occupied, and the experimental valence angle of $104\frac{1}{2}°$ is a compromise between the attempt of the $(1a_1)^2(1b_2)^2$ pair of electrons to increase the valence angle, and of $(2a_1)^2$ to decrease it. This means that if an electron is taken away from the $2a_1$ MO, the valence angle will increase and the molecule become more nearly (or exactly) linear.

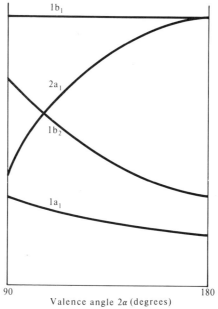

FIG. 46. Walsh diagram for molecules of type AH_2, where A can be any of the atoms Be, B, C, N, or O.

The next step is to suppose that curves similar to those of Fig. 46 apply to all molecules and ions of type AH_2. If so, those molecules with 4 valence electrons (BeH_2, BH_2^+) should be linear, but those with 5, 6, 7, or 8 should be bent. Experiment confirms this view. Thus BH_2, a five-electron molecule, has a valence angle of $131°$ in the ground state, but in the excited state corresponding to excitation $2a_1 \rightarrow 1b_1$ it becomes linear. Six-electron molecules, with two electrons in the $2a_1$ MO, should have a smaller valence angle (CH_2 has $103°$). Seven-electron molecules should also be bent, with about the same angle as six-electron ones. Thus NH_2 has $103°$ for its ground-state valence angle, but it becomes linear in the excitation $2a_1 \rightarrow 1b_1$. Eight-electron molecules, such as H_2O, should have about the same valence angle as those with six or seven electrons (for water, $2\alpha = 104\frac{1}{2}°$).

Diagrams analogous to that in Fig. 46 have been drawn for a wide variety of different structures. Clearly they cannot be used quantitatively unless we know far more than we are likely to know about the shape of the relevant curves. But they have proved remarkably valuable in a qualitative way, and for further details Walsh's original papers should be consulted. It will be obvious that in using concepts of this kind we have abandoned any attempt to represent the molecular binding in terms of two-electron localized bonds. Individual bonds have 'disappeared' in this type of description—a situation which may help us to realize once more than even the concept of a bond is itself only a mental construct: 'bonds' do not exist as separate entities.

Benzene and the world of aromatic molecules

Another example of delocalized orbitals introduces us to the vast world of unsaturated, conjugated, and aromatic molecules. Only the briefest introduction is possible. Benzene is the classic example to choose.

The benzene molecule C_6H_6 is planar; the carbon nuclei form a regular hexagon and the hydrogen atoms lie radially out from each carbon (Fig. 47).

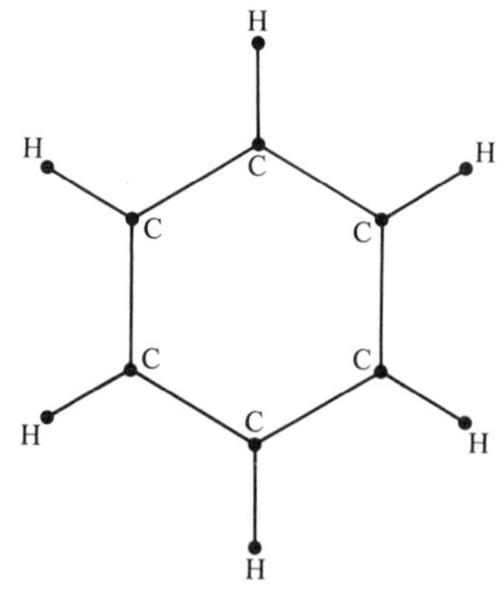

FIG. 47. The benzene molecule C_6H_6.

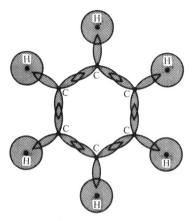

FIG. 48. Formation of C—C and C—H σ-bonds in benzene.

We start with the carbon atoms, and form the sp² trigonal hybrids illustrated in Fig. 27. We can then orient these so that six localized C—C σ-bonds are formed, and also, by putting the six H(1s) orbitals as shown in Fig. 48, we can form six C—H σ-bonds. This leaves us with six electrons. Just as in Fig. 12(d) these electrons occupy AOs of 2p character, whose directions are all parallel, and perpendicular to the plane of the nuclei (Fig. 49(a)). If we wanted to pair these together to form localized π-bonds we could do so as indicated in Fig. 49(b). But there would be nothing to favour this pairing scheme over the alternative pairing scheme in Fig. 49(c). We must use both, and, if we refer to these two pairing schemes as Kekulé structures, we must speak of resonance between them, and use a wavefunction built up from these two (and perhaps other) component structures.

An easier way, however is to extend the idea of delocalized MOs that we saw on p. 72 to be appropriate to H_3^+. Let us make our localized MOs by linear combination of all six of these AOs.† There will in fact be six MOs that

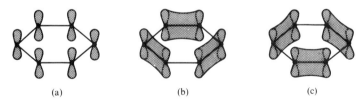

(a)   (b)   (c)

FIG. 49. π-orbitals in benzene. (a) The six identical and parallel atomic orbitals. (b), (c) The two Kekulé pairing schemes.

† This is the LCAO model once more; but now the MOs begin to look like the completely delocalized orbitals for a metallic conductor. Historically it was from this analogy that the LCAO approximation for molecules was devised.

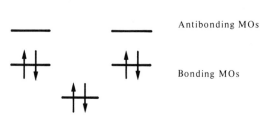

FIG. 50. The molecular energies in the ground state of benzene. Arrows denote π-electrons in the bonding MOs.

can be made from six AOs. Three turn out to be bonding, three are antibonding. As we have just six electrons to put into these MOs we completely fill the bonding ones (Fig. 50).

All six π-electrons now move in orbitals that extend over the whole carbon framework. So any perturbation, such as a substituent attached at one carbon atom, will be able to exert an influence over the whole molecule. This is the fundamental explanation of the *ortho-*, *meta-*, and *para*-directiong effects of substituents in an aromatic molecule.†

This model gives a simple explanation of the great stability of the six-membered ring. For if each carbon atom is to be bonded to two other carbons and one hydrogen, we shall need trigonal hybrids, whose natural valence angle is 120°. Only in a regular hexagon can we obtain this particular value. It is interest to note that rings of this kind with five or seven carbon atoms exist, but in cyclo-octatetraene C_8H_8 the strain in the σ-bonds is so great that the molecule cannot remain planar, and buckles.

It is now very straightforward to extend our description of benzene to other aromatic molecules. For example, in naphthalene ($C_{10}H_8$; see Fig. 51) the valence angles are all close to 120°; there are eleven C—C and eight C—H

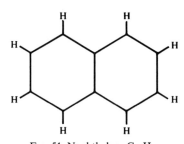

FIG. 51. Naphthalene $C_{10}H_8$.

† A subject discussed by R. A. Jackson in *Mechanism* (OCS 4).

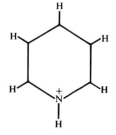

FIG. 52. Pyridine C_5H_5N.

σ-bonds, each using two electrons, and finally the remaining ten π-electrons occupy five bonding MOs, which extend over the whole molecule. The condition for maximum overlapping of neighbouring π atomic orbitals is satisfied if the molecule is planar. Most polynuclear aromatic hydrocarbons are described in an analogous way.

Pyridine (Fig. 52) C_5H_5N is very similar to benzene, in that it is planar and nearly hexagonal. All six atoms of the ring are hybridized in trigonal form: the only significant difference is that two lone-pair electrons, associated with the nitrogen atom, now replace the two electrons in one of the C—H bonds in benzene. Experimental evidence for the existence of these projecting lone-pair electrons (compare Fig. 26 for similar projecting electron-clouds in H_2O) can be found both from the basic property of molecules of this kind, in which the two lone-pair electrons are used to attach a proton and form pyridinium (Fig. 53), and also from X-ray measurements of the total electronic charge density in the molecule. Thus some accurate X-ray studies of s-triazine (Fig. 54), where there are three ring nitrogen atoms, each with its lone pair, show the presence of the projecting charge-cloud very clearly. The difference-density diagram (Fig. 55) reveals both the expected build-up of charge in the σ-bond region between each pair of adjacent ring atoms, and also the concentration of electrons in the three lonepair regions.

FIG. 53. Pyridinium $C_5H_6N^+$.

FIG. 54. s-Triazine $C_3H_3N_3$.

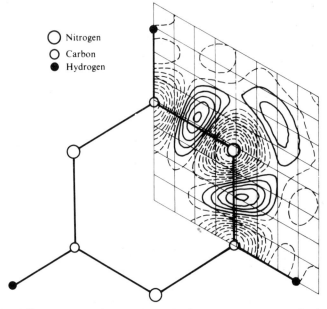

O Nitrogen
O Carbon
● Hydrogen

Fig. 55. Difference density in s-triazine. Solid lines denote an increase in electron density, dotted lines a decrease.

Chemically we know that pyrrole (C_4H_5N; see Fig. 56(a)) is very different from pyridine. We can easily see why this is. If the N atom is trigonally hybridized, and we form the various σ-bonds, we find that the N atom now contributes two π-electrons; in pyridine it contributed only one. Since the N atom cannot accommodate more than two π-electrons, on account of the Pauli exclusion principle, it follows that in pyrrole the net flow of π-electrons must be away from the nitrogen atom; in pyridine, where we start with only one, π-electrons can flow to it. These π-electron migrations, which are superposed upon the charge distribution in the σ-bonds, fit nicely with the measured dipole moments of these two molecules. The internal angles of the

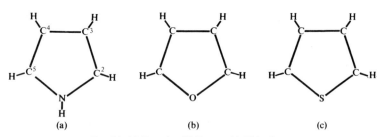

Fig. 56. (a) Pyrrole, (b) Furan, (c) Thiophene.

pyrrole ring are all close to 108°, so that the σ-bond framework is somewhat strained.

The reader will now have no difficulty in understanding the structures of furan (Fig. 56(b)), thiophene (Fig. 56(c)), and other similar types of molecules. It will be recognized that in all these molecules there is a total of six π-electrons associated with each ring. We speak of the *aromatic sextet*. The presence of such a sextet of π-electrons is one condition for good stability. It is clear from Fig. 50 that in benzene the origin of this stability is to be found in the existence of exactly three bonding MOs, each doubly occupied. In all the other cases an MO energy diagram can be drawn, similar to that of Fig. 50 (except that the degeneracy is usually broken). This simple explanation of the aromatic sextet represents one of the most striking contributions of MO theory to the field of aromatic chemistry.

Finally let us return to pyrrole. The two π-electrons on the nitrogen atom are more tightly bound to their nucleus than are the π-electrons of the four carbon atoms. So they will tend to concentrate on the nitrogen. In terms of valence-bond structures, this means that Fig. 57(a) is more favoured than, for example, Fig. 57(b). As a result the carbon bonds are not all equal: C^2—C^3 and C^4—C^5 will have more double-bond character than will C^3—C^4. This is shown in their bond lengths: C^2—C^3 is of length 138 pm and C^3—C^4 is 142 pm. Similar differences exist in other molecules, including pure hydrocarbons such as naphthalene (Fig. 58), and can be described quantitatively using (π-electron) charges and bond orders (p. 39). An extensive literature exists in which experiment and theory are compared, usually with success.†

Transition-metal compounds

As a final example of the value of delocalized orbitals we consider a few of the many compounds of the transition metals. The transition elements form a very wide variety of complexes, ranging from largely ionic compounds such as $[FeF_6]^{3-}$, in which the metal has almost completely lost some of its outermost valence electrons, to hydrated complexes such as $[Ti(H_2O)_6]^{3+}$ in

FIG. 57. Pyrrole. (a), (b) Valence-bond structures. (c) C—C bond lengths (pm).

† See, for example, *CV*, Chapter 8.

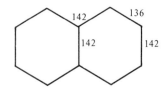

Fɪɢ. 58. Naphthalene bond lengths (pm).

which the bonds have more covalent character. The number of atoms or groups attached to the transition element, usually referred to as ligands, is also widely variable. Such variability of valence is characteristic of atoms which possess an incompletely filled shell of d-orbitals.

In the first transition series† the shell of five 3d AOs is filling as we move from scandium to copper. Table 7 shows the electron configurations of the free atoms in their ground states; the 4s and 3d AOs are of similar size and energy near the beginning of the series but the 3d energy drops, and the 3d orbitals become more compact than the 4s, towards the end. In Fe, for example, the 4s AO is about four times as large as the 3d and the 'outer shell' of s-electrons is readily lost; it is therefore not surprising that in $[FeF_6]^{3-}$ the Fe resembles an ion Fe^{3+}, with only five 3d electrons and no 4s electrons, surrounded by six F^- ions as in Fig. 59. In $[Ti(H_2O)_6]^{3+}$, by contrast, the titanium is better able to keep its share of the valence electrons, even though the complex as a whole carries a positive charge, and the bonding is consequently more covalent in character.

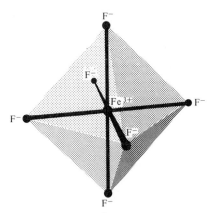

Fɪɢ. 59. Arrangement of ligands in $[FeF_6]^{3-}$. The six fluorine ions lie at the vertices of a regular octahedron, indicated by the light lines.

† Shown in relation to the full periodic table on the front end-paper of the book.

Figure 59 illustrates the simple ionic model which lies at the heart of crystal field theory. Bethe (1929) showed that in the presence of six charged ligands the 3d AOs of the central ion would no longer be degenerate in energy, as they would be in the free ion; and that the pattern of the 'crystal field splitting' would depend on the symmetry of the array of ligands. The more common patterns, with negatively charged ligands, are shown in Fig. 60, the sequence in the centre corresponding to the Fe^{3+} d-orbitals with the octahedral set of ligands in Fig. 59. It is easy to see why the d-orbital energies are split in this way. In Fig. 61 we show two of the d-orbitals used in describing the in-plane bonding in a square planar complex; the other orbitals (d_{z^2}, d_{zx}, d_{yz}) are all normal to the plane and are not shown (cf. Fig. 4). If four negative ligands now approach symmetrically along the x- and y-axes the energy of $d_{x^2-y^2}$ (Fig. 61(a)) will clearly rise considerably because the negative charges move directly towards the four lobes of negative charge, and the repulsion corresponds to an increase of electronic potential energy. The d_{xy} orbital (Fig. 61b)) will clearly be much less strongly affected; and the out-of-plane orbitals still less. The pattern shown on the extreme right of Fig. 60 is thus just what we should expect. The other symmetries can be dealt with in a similar way.

The magnitude of the splittings in Fig. 60 is determined by the nature of the ligands and the strength of the field (the 'crystal field') they produce; this effect, which increases along the 'spectro-chemical series'

$$I^-, \ Br^-, \ Cl^-, \ F^-, \ OH^-, \ H_2O, \ NH_3, \ CN^-, \ CO,$$

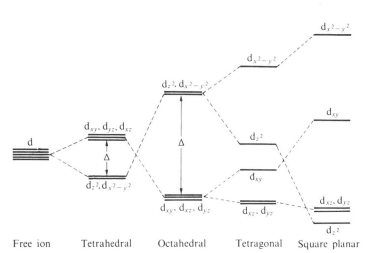

Free ion   Tetrahedral   Octahedral   Tetragonal   Square planar

FIG. 60. Usual order of d-orbital energies. In octahedral symmetry the ligands are on the x-, y-, and z-axes at the centres of six cube faces; in tetrahedral symmetry they are at four cube corners; in tetragonal symmetry the cube is stretched along the z-axis (giving a square prism); in the square planar case the z-axis ligands are removed.

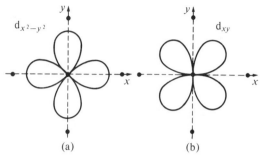

FIG. 61. Square planar complex. The dots represent negatively charged ligands; the lobes contain negative charge.

shows itself in a shift towards the blue of the absorption bands which arise when light is absorbed and an electron of the central metal ion jumps to a higher level. Indeed, the existence of so many brightly coloured salts of the transition metals is essentially due to these crystal field effects.

A few simple deductions can be made at once from Fig. 60. A complex is likely to be stable only when the high-lying, and therefore energetically unfavourable, d-orbitals do not contain electrons: a metal ion such as Ni^{2+}, with 8 d-electrons, would not be stable with four ligands in a tetrahedral array (4 electrons in the highest-lying orbitals) but might well be stable with the square planar arrangement (0 electrons in the highest-lying orbital). And in fact Ni, Pd, and Pt do tend to form square planar complexes, whereas other symmetries are more common when fewer d-electrons are present. Thus chromium and vanadium both form octahedral complexes, $[V(H_2O)_6]^{3+}$ and $[Cr(H_2O)]^{3+}$, in which the central ion (V^{3+} or Cr^{3+}) accommodates 2 or 3 d-electrons, respectively, in the low-lying group of orbitals (Fig. 60, centre column).

To take the crystal field approach further we must ask how the strength of the field will affect the allocation of electrons to the d-orbitals. We know that in the absence of the field the five d-orbitals are degenerate and that by Hund's rules (p. 9) the electrons then tend to occupy them *singly* (with spins parallel) in order to minimize the energy. But what happens when the ligands are present? As the field builds up, the electron configuration and the parallel coupling of the spins both remain initially as they were in the free atom; but a point must come at which the electrons in the higher orbitals will fall into the lower (singly occupied) levels, thus achieving a net reduction of the total energy in spite of the increased repulsion energy due to double occupation. The case of the ferric ion is illustrated in Fig. 62 with six F^- ligands the d-orbitals of the Fe^{3+} remain singly occupied, and there is a resultant total spin of $S = \frac{5}{2}$; but with CN^- ligands, further along the spectrochemical series, the energy falls as electrons drop from the upper to the lower orbitals and the

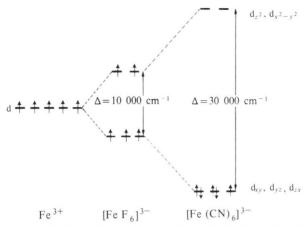

Fe^{3+} $[Fe\,F_6]^{3-}$ $[Fe\,(CN)_6]^{3-}$

FIG. 62. Crystal-field splitting in octahedral complexes of the ferric ion. Small Δ gives a high-spin complex; large Δ gives a low-spin complex.

total spin is also reduced to $S = \frac{1}{2}$. The complexes $[FeF_6]^{3-}$ and $[Fe(CN)^6]^{3-}$ are described as high-spin (weak-field) and low-spin (strong-field) complexes, respectively: and the two types are easily distinguished experimentally through the magnetic effects of the spins—the former being much more strongly paramagnetic than the latter.

It may seem surprising that so much can be done with a classical ionic model, in which there is apparently no 'covalent' bonding of the kind we have associated with electrons in MOs delocalized over at least two atoms. Indeed, why are we treating this topic in a chapter devoted to delocalized bonds? The answer is that the ionic model, though it is crude and oversimplified, leads naturally to a superior approximation in which the electrons are assigned to MOs which are delocalized over the whole complex. To illustrate the main ideas of the resultant 'ligand field theory' we consider one or two examples.

Titanium tetrachloride, $TiCl_4$. Here the outer electron configuration of the neutral metal is $Ti[3d^2 4s^2]$, but in the ionic model of the compound each chlorine would complete its octet, leaving a Ti^{4+} ion surrounded by four Cl^- ions. If we suppose the negative ligands adopt a tetrahedral configuration, Fig. 60 shows that $3d_z^2$ and $3d_{x^2-y^2}$ would be the more tightly bound (lower energy) AOs, while $3d_{xy}, 3d_{yz}, 3d_{zx}$ (and 4s) would be higher in energy and more diffuse. The latter orbitals, according to Table 2, could also be mixed to form four d^3s hybrids pointing towards the four ligands. We may thus pass to an MO description of the bonds by overlapping each d^3s hybrid with a corresponding (doubly occupied) chlorine 2p AO, to obtain a localized MO, or bond orbital, in which the two electrons are to some extent *shared* between the metal and the ligand. In this way the purely ionic picture, which puts an improbably high charge ($+4$) on the metal, is replaced by a more realistic orbital description which allows each bond both covalent and ionic character.

Vanadium tetrachloride, VCl_4. Vanadium has one more valence electron than titanium. The chloride is again tetrahedral and the electronic structure resembles that of $TiCl_4$ except that the extra electron occupies one of the low-energy 3d AOs on the vanadium. Surplus electrons, can thus be accommodated in 'inner' d-orbitals, with *non*-bonding character, instead of having to go into *anti*-bonding MOs; and a given transition metal may in this way form a wide range of stable compounds with varying numbers of non-bonding d-electrons.

Titanium hexachloride ion $[TiCl_6]^{2-}$. Whereas in the tetrachloride the geometry is tetrahedral, the six ligands here form an octahedron. In a pure ionic model the empty d-orbitals of Ti^{4+} would be, from Fig. 60, $3d_{xy}$, $3d_{yz}$, $3d_{zx}$ (tightly bound, inner orbitals) with $3d_{x^2y^2}$, $3d_{z^2}$ lying somewhat higher; the latter could mix with 4s and 4p to give (Table 2) six d^2sp^3 hybrids pointing directly towards the ligands. The rest of the argument is exactly the same as for the tetrachloride (above).

Ferricyanide ion $[Fe(CN)_6]^{3-}$. The geometry is again octahedral and similar considerations apply. In the ionic model the metal would be regarded as $Fe^{3+}[3d^5]$ surrounded by six CN^- ions: the five d-electrons could be accommodated in the 'inner' shell (d_{xy}, d_{yz}, d_{zx}), leaving $d_{x^2-y^2}$ and d_{z^2} available for bonding. Again d^2sp^3 hybridization would be appropriate and would permit the construction (by overlap with carbon lone-pair AOs on the CN groups) of localized MOs which would introduce some covalent character into the bonds and allow charge to flow back towards the metal.

The above examples illustrate Pauling's approach[†] to the bonding in transition metal compounds, which is based on the use of hybridization. Ligand field theory in its more complete form, however, does not attempt to recognize localized electron-pair bonds; instead, it accepts from the start a completely *de*localized description of the bonding, building up the MOs for the whole complex from all the available AOs—just as we have done earlier in this chapter in dealing with conjugated systems.

We can do no more than hint at the more flexible form of ligand field theory.[‡] Let us consider a square planar system but, instead of looking for suitable hybrids, arrange the AOs on the metal and its ligands into various 'symmetry combinations'. Examples of such combinations of ligand orbitals, showing the metal d-orbitals with which they interact, are shown in Fig. 63.

† L. Pauling (1931). *J. Am. Chem. Soc.* **53**, 1367.
‡ Further details are given in *CV*, Chapter 9.

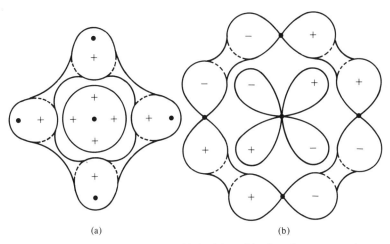

(a)   (b)

FIG. 63. Formation of ligand group orbitals: (a) combination of same symmetry as central As AO; (b) combination of same symmetry as central $3d_{xy}$ AO. (Broken lines indicate the ligand orbitals used.)

Each set of ligand orbitals, put together with coefficients chosen to give the distribution of signs indicated, is a 'ligand group orbital'—essentially a 'symmetry orbital' whose behaviour under symmetry operators (e.g. reflections across planes of symmetry) mimics that of an AO on the central atom. MOs of each given 'symmetry species' are then often well approximated by pairwise combination of metal orbitals and group orbitals belonging by pairwise combination of metal orbitals and group orbitals belonging to the same species; and the bonding may be discussed, using for example correlation diagrams (cf. Fig. 18), in much the same way as for a simple diatomic molecule. The bonding is truly delocalized, as Fig. 63 shows, and the separation of the MOs into symmetry types is rather more complicated; but it is still possible to recognize essentially σ- and π-type bonding (Figs, 63(a) and (b)) of the kind we encountered in Chapter 2; and many basic concepts—such as the maximum overlap principle—can be carried over with little change.

With this glimpse of two vast fields of application—to organic chemistry and to inorganic chemistry—we must end this outline of the orbital approach. We have tried to show how a few quite simple ideas, not particularly abstruse or mathematical, have allowed us to rationalize a large part of structural chemistry. Our discussion has taken the classical symbol for a chemical bond—the straight line of Archibald Couper—and given it a pictorial character. Even if, at this stage, we are not always able to predict in advance what should be the size and shape of a molecule, we can sometimes do so; and always we have a deeper understanding of why molecules are what they are.

Further reading

Chapter 1

A very complete listing of experimentally determined molecular sizes and shapes is given in *Tables of Interatomic Distances and Configurations in Molecules* (1958), ed. by Sutton, L. E., Special Publication no. 11. London: the Chemical Society: and in the (1965) Supplement to the above: Special Publication no. 18. London: the Chemical Society.
An elementary account of the wave mechanics necessary for this book is to be found in McWeeny, R. (1979) *Coulson's Valence*, Clarendon Press, Oxford (the revised 3rd edition of Coulson, C. A. *Valence*). This work is referred to as *CV* throughout the present book. Reference may also be made to Atkins, P. W. (1983) *Molecular Quantum Mechanics*, second edition, Clarendon Press, Oxford. A discussion of the Born–Oppenheimer approximation has been given by Teller, E. and Sahlin, H. L. (1970) in *Physical Chemistry* Volume 5 (Valency) eds. Eyring, Henderson, and Jost, Academic Press, New York and London.
Results of *ab initio* calculations on molecules are listed in a series of books by Richards, W. G. and co-workers (*A Bibliography of ab initio Molecular Wave Functions*, Clarendon Press, Oxford, 1971, and subsequent supplementary volumes).

Chapter 2

A table of many of the most important wave functions for H_2, using MO and VB approximations, is given in *CV*, Chapter 5. The equivalence of MO and VB wave functions when appropriate extensions of each have been made, was shown in detail for H_2 by Coulson, C. A. *and* Fischer (Miss) I. (1949), *Phil. Mag.* **40**, 386; and is also discussed fully in *CV*, Chapter 5. Some of the difficulties inherent in the calculation of molecular dipole moments, taking account of the contribution from the other non-bonding electrons and of other effects, are set out in *CV*, Chapter 6, along with a discussion of the electronegativity concept and of fractional ionic character.

Chapter 3

The use of hybrid orbitals goes back to Linus Pauling (1931) *J. Am. chem. Soc.*, **53**, 1367. The idea of the Valence State is due to van Vleck, J. H. (1933) *J. chem. Phys*, **1**, 177, 219; (1934), ibid, **2**, 20. Extensions are by Mulliken, R. S. (1934) *J. chem. Phys.* **2**, 782; by Moffit, W. E. (1950) *Proc. R. Soc.* A **202**, 534, 548, and by other later writers. The importance of the overlap integral was first stressed by Maccoll, A. (1950) *Trans. Faraday Soc.*, **46**. 369; and bent bonds were introduced by Coulson, C. A. *and* Moffitt, W. E. (1949) *Phil. Mag.*, **40**, 1. Further information about the topics discussed in this chapter may be found in *CV*, Chapter 7, and in an article by Coulson, C. A. (1970) in *Physical Chemistry* Volume 5 (Valency) eds Eyring, Henderson, and Jost, Academic Press, New York and London, Chapter 6.

Chapter 4

Further properties of atomic hybrids are discussed in *CV*, Chapter 7. Details of the sizes and shapes of many of the molecules discussed in this chapter are in the two Chemical Society Special Publications listed in Chapter 1. Other information is in Wells, A. F. (1983), *Structural Inorganic Chemistry*, Clarendon Press, Oxford, 5th edition. An elementary survey of the rare-gas compounds is in Coulson, C. A. (1964), *J. chem. Soc.*, 1442. The interpretation of shape in terms of Walsh's rules is discussed by Coulson, C. A. (1970) in Chapter 6 of *Physical Chemistry* Volume 5 (Valency), cited above. An alternative treatment of polyatomic molecules, which stresses the repulsion between lone-pair electrons and all other electrons, along the lines introduced by Sidgwick, Powell, Nyholm, and Gillespie, may be found in the book *Molecular geometry* by R. J. Gillespie, Van Nostrand (1972).

Chapter 5

The major early developments in delocalized MOs are due to R. S. Mulliken; for an early review see Mulliken, R. S. (1932), *Rev. mod. Phys.* 4, 1. Many references to recent calculations are given in Mulliken, R. S. and Ermler, W. C. (1977) *Diatomic Molecules: Results of Ab Initio Calculations* and (1981) *Polyatomic Molecules: Results of Ab Initio Calculation*, Academic Press, New York and London. A convenient set of group-theory tables, useful for classifying the MOs in water and other molecules, is *Tables for Group Theory*, by Atkins, P. W., Child, M. S., and Phillips, C. S. G. (1970), Oxford, Clarendon Press. A full account of the many applications of MO theory to aromatic and unsaturated molecules is in the book *The Molecular-Orbital Theory of Conjugated Systems*, by Salem, L. (1966), W. A. Benjamin, New York. A simple account is in the book *Molecular-Orbital Theory for Organic Chemists*, by Streitwieser, A. (1961), John Wiley, New York and London. See also *CV* Chapter 8 and Coulson, C. A. (1970) in Chapter 7 of *Physical Chemistry*, Volume 5 (Valency), cited above. For delocalized bonding in the boron hydrides see Lipscomb, W. N. (1963) *Boron Hydrides*, Benjamin, New York, and for transition-metal compounds see Orgel, L. E. (1966) *An Introduction to Transition-Metal Chemistry* (2nd ed.), Methuen, London, or for a brief account *CV*, Chapter 9.

Related books in the Oxford Chemistry Series

ATKINS, P. W. (1974) *Quanta: a handbook of concepts.*
JACKSON, R. A. (1972) *Mechanism: An introduction to the study of organic reactions.*
McLAUCHLAN, K. A. (1972) *Magnetic resonance.*
PASS, G. (1973) *Ions in solution (3): Inorganic properties.*
PUDDEPHATT, R. J. (1972) *The periodic table of the elements.*
WORMALD, J. (1973) *Diffraction methods.*

Index

ab initio calculations, 16, 17
alkali atoms, valence behaviour, 60
ammonia, 64
borane, NH_3 . BH_3, 68
ammonium NH_4^+, 64
anthracene, 2
electronic density contours, 2
antibonding molecular orbitals, 32
antisymmetrization, 8
antisymmetry and spin, 24
aromatic molecules, 76
sextet, 80
atomic orbitals, 3
normalization, 7
orthogonality, 7
s, p, d type, 4–6
shells, 4
size, 4, 23
aufbau procedure, 6–10, 32
axial symmetry of molecular orbitals, 32

B atom, 9
BF_4^-, 68
BH, 42
BH_2^+, 76
BH_4^-, 68
B_2H_6, 73
Baeyer theory of strain, 53
Be atom, 9
BeH, 41
BeH_2, 76
BeO, 66
bent bonds, 51
benzene, nuclear density contours, 3
electronic structure, 76
molecular shape, 50
Bohr radius, 4
bond order, 39, 81
bonds,
bent, 51
dative, 59
delocalized, 72–87
donor, 26, 59,
double, 26–7, 50
localized, 44
properties, 44
three-centre, 73
three-electron, 70
triple, 27, 50
bonding molecular orbitals, 32

Born–Oppenheimer approximation, 11, 12

C atom, 9, 48
covalent radius and hybridization, 69
valence states, 49
CF_2, 49
carbon–hydrogen bonds, 69
CH radical, 42
CH_2, 49
CH_4, 12, 49
CO_2, 11
charges (populations), 39, 81
charge-cloud, 1, 3
density, 4, 22, 38–40
CIF_3, 62
closed-shell octet, 68
complexes, transition metal, 81
chromium, 84
ferric, 81–5
high-spin (weak-field), 85
low-spin (strong-field), 85
titanium, 81–2, 85, 86
vanadium, 84, 86
contour diagram, 2, 22
covalent function, 23, 28
covalent radius and hybridization, 69
crystal field, theory, 83–5
splitting, 83
CsF, 60
cyclo-octatetraene, 78
cyclopropane, 51
tricyano-, 52

d-orbital, 6
dative bond, 59
degeneracy, 4
delocalized bonds, 72–87
δ bond, 33
Δ-states, 33
diamond, 40
diatomic molecules, 19–43
diborane, B_2H_6, 73
difference density, 39, 40
diffraction,
electron, 2
neutron, 2
X-ray, 2
digonal hybrids, 51
dipole moment, 29, 36, 41, 79

directing effect, of substituents, 77
donor bonds, 59
double bond, 26–7, 50
character, 81
doublet state, 7

eigenfunction, 13
eigenvalue, 13
electron configuration, 9, 10, 33, 34
diffraction, 2
pair, bond, 21
wave function, 21
spin, 7
resonance, 2
electronegativity, 28, 30
ethene (ethylene), 50, 69
ethyne (acetylene), 51
exchange, 10, 22, 54
exclusion, see Pauli principle

fluorine, atom, 9
molecule, 25, 34
fractional ionic character, 29
formaldehyde, 68
furan, 80

gerade symmetry, 33

H atom, 9
molecule, 17, 20, 30
charge density, 22, 38
difference density, 40
H_3, 12
H_3^+, 72
H_2O, see water
H_3O^+, 64
halogens, 61
Hamiltonian, 13
for helium, 14
Heitler–London wave function for H_2, 20–2
refinements, 23
helium, atom, 9, 14, 17
molecule, 35, 70
heteronuclear diatomics, 28, 35
HF, 28, 37, 41
$HgMe_2$, 67
homonuclear diatomics, 25, 32
Hund's rules, 9
hybridization, 46
advantages and disadvantages, 53
and covalent radius, 69
different types, 56–7
digonal, 51
tetrahedral, 49

trigonal, 50
and valence angles, 47
hydrogen bond, 48
hydrogen halides, 29
hydrazine, 65

inversion symmetry, 32
ionic character, 29
ionic function, or structure, 23, 28, 41
ionic model (complexes), 85
ionization potential, 28, 35

Kekulé structures, 77
KrF_2, 70

LCAO approximation, 36, 77
Li, atom, 9
molecule, 25, 34, 46, 60
difference density, 41
molecule ion, 60
ligand field theory, 85–7
ligand (group) orbitals, 86, 87
ligands, 82, 83
LiH, 41
localized charge distribution, 44
lone-pair electrons, 48, 65, 78
mutual repulsion, 65

many-centre bonds, 72
maximum overlapping, 25
Mendeleev, see periodic table
methane, 12, 48
molecules, aromatic, 76
heteronuclear diatomics, 28, 35
homonuclear diatomics, 25, 32
polyatomics, 44–58
molecular-orbital method, 19, 30
molecular orbitals, 30
δ-type, 33
for homonuclear diatomics, 32–5
for HF, 38
π-type, 33
σ-type, 30, 32

NaCl, diatomic, 60
napthalene, 79
neon, atom, 9
molecule, 34
neutron diffraction, 2
for benzene, 3
nitrogen, atom, 9
molecule, 27, 34, 35
NH radical, 42
NH_2 radical, 76
nuclear positions, 2, 3
nuclear spin coupling, 2

o-, m-, p-directing power (substituents), 77
octet, 68
orbitals, atomic, 3
    delocalized, 72–87
    localized, 44
    molecular, 30
    orbital model, 3
    orbital energy, 8
    diagram, 10
orthogonality, 7, 47
overlap, integral, 7, 25, 46, 55
    region, 25, 38–9
    π-type, 27
    σ-type, 27
oxygen, atom, 9
    molecule, 34
OH radical, 42
oxonium compounds, 64

Pauli principle, 7–8, 22, 24
p-orbital, 5
paramagnetism, of O_2, 34
    of complexes, 85
PCl_5, 65
percentage ionic character, 29
periodic table, 8, 59
π-bond, 26, 33
Π-states, 33
± symmetry, 32
polyatomic molecules, 44–58
polyhalogenides, 62
populations, 39
potential-energy, curve, 11, 12, 19, 21
    surface, 12
principal quantum number, 5
principle of maximum overlapping, 25
pyridine, 78
pyridinium, 79, 80
pyrrole, 79, 81

quantum chemistry, 16
quantum numbers for atoms, 4

rare-gas atoms, 69
Rayleigh's, principle, 13
    ratio, 14
reflection symmetry (in MOs), 32
repulsive states, 21
resonance, 23, 70, 77
restricted rotation, 26, 51
Ritz method, 15, 23, 31

s-bond, 60
s-orbital, 4

s-triazine, 79, 80
Schrödinger wave equation, 13
semi-empirical calculations, 17
separated-atom viewpoint, 19, 20, 30
sextet, aromatic, 80
SF_6, 64
shared electrons, 22
shells, (K, L, M,), 4
σ-bond, 26, 32
Σ-states, 33
single bond, 26
singlet state, 24
Slater-type orbital, 15
sodium atom, 9
spectrochemical series, 83
spin, 7–8, 24
spin-orbital, 8, 24
strain, 51
structure of molecules, 1
substituent effects in benzene, 77
symmetry combinations (orbitals), 86–7
symmetry properties, 32

tellurium, 63
$TeCl_4$, 64
tetrahedral, carbon atom, 49
    hybrids, 49
    for B, N, etc., 68
thiophene, 80
transition elements, 59, 81
transition series, 82
transition-metal compounds, 81
trigonal hybrids, 50
triple bond, 27, 50
triplet state, 24
ungerade symmetry, 33
united-atom viewpoint, 19, 30

valence angles,
    barium, 67
    group II, 66
    group III, 68
    group IV, 68
    group V, 64
    group VI, 63
    mercury, 67
    rare-gas compounds, 70
    water, 45, 47, 76
valence number, 59
valence rules, 59–71
valence state, 49, 59
valences, higher, 62, 64
valence-bond function 21, 59
valence-bond method, 19

variation method, 13
    for atoms, 14
    for molecules, 23
vibrations of molecules, 1
    amplitudes, 1
VSEPR theory, 54

Walsh diagram, 74
water molecule, 11, 44, 47, 63, 74, 76

xenon fluorides, 70, 72
X-ray diffraction, 2

IA	IIA	IIIA	IVA	VA	VIA	VIIA	VIII			IB	IIB	IIIB	IVB	VB	VIB	VIIB	O
$_1$H 1·008																	$_2$He 4·003
$_3$Li 6·941	$_4$Be 9·012											$_5$B 10·81	$_6$C 12·01	$_7$N 14·01	$_8$O 16·00	$_9$F 19·00	$_{10}$Ne 20·18
$_{11}$Na 22·99	$_{12}$Mg 24·31											$_{13}$Al 26·98	$_{14}$Si 28·09	$_{15}$P 30·97	$_{16}$S 32·06	$_{17}$Cl 35·45	$_{18}$Ar 39·95
$_{19}$K 39·10	$_{20}$Ca 40·08	$_{21}$Sc 44·96	$_{22}$Ti 47·90	$_{23}$V 50·94	$_{24}$Cr 52·00	$_{25}$Mn 54·94	$_{26}$Fe 55·85	$_{27}$Co 58·93	$_{28}$Ni 58·71	$_{29}$Cu 63·55	$_{30}$Zn 65·37	$_{31}$Ga 69·72	$_{32}$Ge 72·59	$_{33}$As 74·92	$_{34}$Se 78·96	$_{35}$Br 79·90	$_{36}$Kr 83·80
$_{37}$Rb 85·47	$_{38}$Sr 87·62	$_{39}$Y 88·91	$_{40}$Zr 91·22	$_{41}$Nb 92·91	$_{42}$Mo 95·94	$_{43}$Tc 98·91	$_{44}$Ru 101·1	$_{45}$Rh 102·9	$_{46}$Pd 106·4	$_{47}$Ag 107·9	$_{48}$Cd 112·4	$_{49}$In 114·8	$_{50}$Sn 118·7	$_{51}$Sb 121·8	$_{52}$Te 127·6	$_{53}$I 126·9	$_{54}$Xe 131·3
$_{55}$Cs 132·9	$_{56}$Ba 137·3	$_{57}$La 138·9	$_{72}$Hf 178·5	$_{73}$Ta 180·9	$_{74}$W 183·9	$_{75}$Re 186·2	$_{76}$Os 190·2	$_{77}$Ir 192·2	$_{78}$Pt 195·1	$_{79}$Au 197·0	$_{80}$Hg 200·6	$_{81}$Tl 204·4	$_{82}$Pb 207·2	$_{83}$Bi 209·0	$_{84}$Po (210)	$_{85}$At (210)	$_{86}$Rn (222)
$_{87}$Fr (223)	$_{88}$Ra 226·0	$_{89}$Ac (227)															

Lanthanides	$_{57}$La 138·9	$_{58}$Ce 140·1	$_{59}$Pr 140·9	$_{60}$Nd 144·2	$_{61}$Pm (147)	$_{62}$Sm 150·4	$_{63}$Eu 152·0	$_{64}$Gd 157·3	$_{65}$Tb 158·9	$_{66}$Dy 162·5	$_{67}$Ho 164·9	$_{68}$Er 167·3	$_{69}$Tm 168·9	$_{70}$Yb 173·0	$_{71}$Lu 175·0
Actinides	$_{89}$Ac (227)	$_{90}$Th 232·0	$_{91}$Pa 231·0	$_{92}$U 238·0	$_{93}$Np 237·0	$_{94}$Pu (242)	$_{95}$Am (243)	$_{96}$Cm (248)	$_{97}$Bk (247)	$_{98}$Cf (251)	$_{99}$Es (254)	$_{100}$Fm (253)	$_{101}$Md (256)	$_{102}$No (254)	$_{103}$Lw (257)